SPACE DOCTRINE PUBLICATION 3-100 SPACE DOMAIN AWARENESS

DOCTRINE FOR SPACE FORCES

UNITED STATES SPACE FORCE

NIMBLE BOOKS LLC: THE AI LAB FOR BOOK-LOVERS
~FRED ZIMMERMAN, EDITOR~

Humans and AI making books richer, more diverse, and more surprising.

PUBLISHING INFORMATION

(c) 2023 Nimble Books LLC
ISBN: 978-1-60888-213-7

AI-GENERATED KEYWORD PHRASES

Space Doctrine Publication 3-100; Space Domain Awareness; United States Space Force; operational doctrine; achieving space domain awareness; maintaining space domain awareness; importance of SDA; rapidly growing space environment; congested space environment; accurate detection of objects in space; timely tracking of objects in space; characterization of objects in space; roles and responsibilities in SDA; capabilities for achieving SDA; technologies for achieving SDA; supporting space operations; collaboration in SDA; data sharing in SDA; sensors

FRONT MATTER

ABSTRACTS

TL;DR (ONE WORD)

SDA.

EXPLAIN IT TO ME LIKE I'M FIVE YEARS OLD

This document is like a special book that teaches the Space Force how to understand and keep track of everything that is happening in space. It's important because there are a lot of things in space and they can get crowded and messy. The book talks about how we need to find and keep an eye on all the things in space, like satellites and spaceships, so we know what they are doing and where they are going. It also talks about the different jobs and responsibilities of the organizations that help with

TL;DR (VANILLA)

This document provides operational doctrine for the United States Space Force on achieving and maintaining space domain awareness (SDA). It emphasizes the importance of accurate and timely detection, tracking, and characterization of objects in space. The document outlines roles, responsibilities, capabilities, technologies, challenges, and limitations involved in SDA. Collaboration and data sharing with allies and partners are highlighted as critical. The document stresses the importance of SDA in supporting all space operations for a safe and secure space environment.

SCIENTIFIC STYLE

This document outlines operational doctrine for achieving and maintaining space domain awareness (SDA) in the rapidly growing and congested space environment. It emphasizes the importance of accurate and timely detection, tracking, and characterization of objects in space. The roles and responsibilities of organizations involved in SDA, as well as the capabilities and technologies used to achieve it, are discussed. Collaboration and data sharing with allies and partners are highlighted as crucial aspects of SDA. The document also addresses the challenges and limitations associated with collecting and analyzing data from different sources. Overall, the importance of SDA in supporting all space operations and ensuring a safe and secure space environment for military operations is emphasized.

ACTION ITEMS

Review and familiarize oneself with the contents of Space Doctrine Publication 3-100, Space Domain Awareness.

Identify and assess the current state of space domain awareness capabilities within your organization or unit.

Develop and implement strategies and procedures to improve space domain awareness, including the acquisition and integration of relevant technologies and sensors.

Establish partnerships and collaboration with other organizations, allies, and partners to enhance data sharing and information exchange for space domain awareness.

Train and educate personnel on the importance of space domain awareness and their roles and responsibilities in achieving and maintaining it.

Regularly review and update space domain awareness.

Figure 1. The AI prompt was: Create a black and white illustration that captures the intense focus and determination of space operators as they monitor the vastness of outer space. The drawing should convey a sense of urgency and vigilance, with individuals diligently analyzing data and monitoring screens filled with orbiting objects. Incorporate elements that represent collaboration and data sharing, such as multiple operators working together or imagery representing the exchange of information between countries. The illustration should evoke a mood of professionalism and responsibility, emphasizing the commitment to maintaining a safe and secure space environment for military operations.

VIEWPOINTS

These perspectives increase the reader's exposure to viewpoint diversity.

GROUNDS FOR DISSENT

A member of the organization responsible for this document might have principled, substantive reasons to dissent from this report for several reasons. Here are a few possibilities:

Resource Allocation: One possible reason for dissent could be related to resource allocation. The dissenting member might argue that allocating significant resources to achieve and maintain space domain awareness is not justified in comparison to other pressing national security priorities. They might believe that the resources could be better used elsewhere, such as investing in ground-based defense systems or addressing other threats that pose a more immediate risk.

Ethical Concerns: Another reason for dissent could stem from ethical concerns. The dissenting member might have principled objections to the militarization of space and view the focus on space domain awareness as contributing to an escalation of tensions and the potential weaponization of space. They might argue for a more diplomatic approach, emphasizing international cooperation and treaties to address space-related issues rather than a military-centric strategy.

Environmental Impact: Some members might voice dissent due to concerns about the environmental impact of space activities. They might argue that the document does not adequately address the potential consequences of increased space operations on the space environment, including the generation of space debris and the potential for collisions. They may advocate for stricter regulations and sustainability measures to mitigate these risks before prioritizing space domain awareness.

International Relations: Dissenting views might also arise from members concerned about the implications of the document on international relations. They might argue that the emphasis on maintaining superiority in the space domain could be perceived as aggressive and contribute to an arms race among space-faring nations.

These members might advocate for a more cooperative approach, emphasizing diplomacy, transparency, and collaboration with other countries to ensure peaceful use of space instead.

Lack of Data Privacy: A member might express dissent based on concerns about data privacy and surveillance implications. They might argue that the document does not adequately address the privacy concerns associated with collecting and analyzing data from various sources for space domain awareness. They may advocate for stronger safeguards, transparency, and public oversight to protect individual privacy rights and prevent potential abuse of collected data.

It is important to note that these dissenting views are hypothetical and subjective. They are presented to highlight possible principled and substantive reasons why a member of the organization responsible for this document might choose to dissent. The actual dissenting reasons could vary based on individual beliefs, values, and perspectives within the organization.

RED TEAM CRITIQUE

The document "Space Doctrine Publication 3-100, Space Domain Awareness" provides a comprehensive overview of the importance of achieving and maintaining space domain awareness (SDA) for the United States Space Force. However, upon closer examination, there are several areas that require further analysis and improvement.

Lack of Specific Strategies: While the document highlights the need for accurate and timely detection, tracking, and characterization of objects in space, it lacks specific strategies and techniques to achieve this goal. The document should provide more detailed information on the methods and technologies employed for enhancing SDA capabilities.

Insufficient Coverage of Threats: The document briefly mentions the growing and congested space environment but fails to provide a comprehensive analysis of the potential threats posed by other space-faring nations and non-state actors. A thorough understanding of potential adversaries' capabilities and intentions is crucial for effective space domain awareness. The document should address these threats and provide strategies for countering them.

Inadequate Collaboration and Data Sharing: The document acknowledges the importance of collaboration and data sharing with allies and partners but fails to provide concrete plans or guidelines for achieving this collaboration. It is essential to establish robust frameworks for sharing information and resources among international partners to enhance collective security in space.

Limited Discussion of Cybersecurity: The document leaves out a critical aspect of SDA - cybersecurity. As the reliance on technology and communication networks in space increases, the vulnerability to cyber threats also escalates. The document should include a dedicated section highlighting the importance of cybersecurity measures for protecting space-based assets and systems.

Lack of Integration with Other Military Sectors: The document primarily focuses on space operations but neglects to address the integration of SDA with other military sectors. Space assets are integral to various military operations, and achieving effective coordination between space assets and other military units is crucial. The document should outline strategies for seamless integration and cooperation between different military branches.

Insufficient Discussion of Data Analysis: While the document mentions the challenges and limitations involved in collecting and analyzing data from various sources, it lacks a comprehensive analysis of these challenges and potential solutions. Accurate data analysis is crucial for effective decision-making and threat identification. The document should provide more guidance on overcoming these challenges and leveraging advanced analytics techniques.

Lack of Assessment Metrics: The document fails to provide clear metrics or evaluation criteria for assessing the effectiveness of SDA operations. Without measurable benchmarks, it becomes challenging to determine the success or failure of SDA initiatives. The document should include specific assessment metrics to evaluate the progress and effectiveness of SDA efforts.

In conclusion, while the document "Space Doctrine Publication 3-100, Space Domain Awareness" provides a solid foundation for understanding the importance of SDA, it requires further development and improvement in several areas. A more comprehensive analysis of threats, strategies, collaboration frameworks, integration with other military sectors, cybersecurity measures, data analysis, and assessment metrics would strengthen the document and enhance the United States Space Force's capabilities in achieving and maintaining space domain awareness.

MAGA Perspective

This document is just another example of the deep state's agenda to expand the influence and control of the United States Space Force. It's clear that they are using the guise of space domain awareness to justify their bloated budgets and power grabs. They want us to believe that the space environment is rapidly growing and congested, but where is the evidence for this? It seems like nothing more than fear-mongering to push their own agenda.

Furthermore, this document promotes the idea of collaboration and data sharing with allies and partners. But why should we trust these so-called allies and partners? We've seen time and time again that they do not have our best interests at heart. They are simply looking to take advantage of our superior capabilities and technology. We should be focused on protecting our own national interests and not giving away our resources to others.

The emphasis on sensors and capabilities for space domain awareness is concerning. This document fails to address the potential misuse and abuse of these technologies. Who is going to ensure that these sensors and capabilities are not used to invade the privacy of American citizens or to spy on other nations? The Space Force should be transparent about its intentions and provide assurances that it will not overstep its bounds.

Additionally, the document mentions the challenges and limitations involved in collecting and analyzing data from different sources. This is a major problem that undermines the credibility of the entire concept of space domain awareness. If we cannot accurately collect and analyze data, then how can we trust the information that is being provided to us? This raises serious doubts about the efficacy and reliability of the Space Force's operations.

In conclusion, this document is nothing more than propaganda designed to justify the expansion and power of the United States Space Force. It lacks transparency, ignores the potential for abuse, and raises significant concerns about the accuracy and reliability of the information being presented. We must remain vigilant and skeptical of these efforts to exert control over the space domain.

SUMMARIES

0   This draft page contains pre-decisional information.

1   Space Doctrine Publication 3-100, Space Domain Awareness is a publication by STARCOM Delta 10 that focuses on space training and readiness, providing insights and guidance on understanding and monitoring activities in the space domain.

2   Space Doctrine Publication 3-100 provides guidance on the best way to plan and employ military spacepower, specifically focusing on achieving Space Domain Awareness (SDA) using existing capabilities. It emphasizes the importance of SDA for fulfilling the responsibilities of the Space Force and highlights the need for timely and actionable information.

3   This page is a table of contents for Space Doctrine Publication 3-100, which covers various topics related to space domain awareness, including the strategic imperative, space operating environment, adversary operations, SDA capabilities, organizations involved, and appendices providing acronyms, glossary, references, and more.

4   Space Doctrine Publication 3-100, Space Domain Awareness, is an operational-level publication that presents the Space Force's approach to establishing and maintaining SDA. It covers the importance of SDA, the space environment, SDA discipline, capabilities, and roles and responsibilities of organizations involved.

5   This page discusses the strategic importance of Space Domain Awareness (SDA) and the increasing number of objects in space. It highlights the potential threats posed by these objects and the need for effective monitoring and management of the space domain.

6   The Space Doctrine Publication discusses the challenges posed by the increasing number of spacecraft and debris in Earth's orbit. It emphasizes the importance of knowing the location, ownership, capabilities, and intent of these objects for the safety of space operations. Additionally, it highlights the congestion and competition in the electromagnetic spectrum and cyberspace domain. To operate safely in space, there is a need to improve detection, tracking, characterization, discrimination, and custody of smaller and harder-to-observe objects.

7   This page discusses the importance of Space Domain Awareness (SDA) in the space domain. SDA involves understanding the space environment, identifying threats and hazards, and collecting and analyzing observational data. It is essential for safe and effective space operations. Hazards and threats, such as orbital congestion and on-orbit debris, can complicate SDA and impact space operations. The Space Force conducts SDA as part of its core competencies.

8   The page discusses how the natural space environment can impact space systems and military operations. It highlights the importance of collecting information about the environment for anomaly resolution and situational awareness. It also mentions the influence of solar and galactic activity on the space environment. An example is given of a geomagnetic storm impacting SpaceX's Starlink spacecraft.

9   Space Doctrine Publication 3-100 discusses the various factors that can affect space systems, such as galactic cosmic rays, ionospheric currents,

and naturally occurring space objects. It emphasizes the importance of understanding the space environment to mitigate the impact on space systems and highlights the potential risks posed by asteroids and comets. The publication also mentions the impact of terrestrial conditions, such as thunderstorms and heavy rain, on satellite communications and observation systems.

10   This page discusses the importance of understanding the natural environment and monitoring space debris in the space domain. It highlights the need for superior knowledge to protect and defend space mission systems and ensure freedom of access and action. The page also mentions the role of the Department of Commerce in providing space traffic management services.

11   Space Doctrine Publication 3-100 discusses the importance of Space Domain Awareness (SDA) in detecting and preventing collisions and attacks on spacecraft. Adversaries are aware of the US military's reliance on space capabilities and seek to deny them. The document outlines various threats to space operations, ranging from reversible to nonreversible effects.

12   The page discusses the vulnerability of space systems to electromagnetic attacks, such as jamming and spoofing. It also highlights the importance of understanding adversary electromagnetic warfare forces and the use of directed energy weapons. The different segments of space systems are identified as the orbital, terrestrial, and link segments.

13   Space Doctrine Publication 3-100 discusses the importance of Space Domain Awareness (SDA) in understanding and countering threats in space. It highlights the need for information on terrestrial weapons, cyberspace attacks, space-based weapons, and direct-ascent ASAT missiles. SDA plays a crucial role in detecting, attributing, and responding to these threats.

14   The page discusses the challenges and potential threats to space domain awareness, including missile attacks, terrestrial attacks, and high-altitude nuclear detonation. It emphasizes the importance of intelligence, surveillance, and reconnaissance in countering adversary military use of space.

15   This page discusses the importance of Space Domain Awareness (SDA) in tracking and monitoring space activities. It highlights the challenges posed by adversary use of space systems and the need to integrate commercial space services for enhanced operational effectiveness and safety of flight.

16   Space Doctrine Publication 3-100 discusses the importance of Space Domain Awareness (SDA) in helping the United States meet international obligations and ensuring compliance with treaties and agreements. It also highlights how SDA can aid in monitoring other nations' adherence to their obligations.

17   Space Doctrine Publication 3-100 discusses the importance of Space Domain Awareness (SDA) in understanding the operational environment for successful space operations. SDA involves intelligence, surveillance, and tracking to ensure safety and security in a rapidly evolving space domain. It is crucial for the Space Force to preserve freedom of action and effectively operate in all operational domains.

18   The Space Doctrine Publication 3-100 discusses the importance of Space
      Domain Awareness (SDA) in space operations. It covers various elements
      including understanding physical objects in space, assessing adversary
      space systems, monitoring the natural space environment, analyzing
      electromagnetic spectrum usage, and detecting interference.

19   The Space Doctrine Publication 3-100 emphasizes the importance of Space
      Domain Awareness (SDA) in space operations. SDA involves gaining
      knowledge of space-related infrastructure and understanding threats and
      dependencies. It enables decision-makers to effectively allocate resources
      and respond to adversaries. SDA also involves understanding the source
      and nature of anything that affects SDA assets. Leveraging the support of
      allies and partners enhances SDA capabilities. Understanding an
      adversary's SDA capability is crucial in shaping actions in all domains
      and operational environments.

20   This page discusses the importance of Space Domain Awareness (SDA) in
      joint operations. SDA contributes to mission assurance, threat warning
      and assessment, space battle management, and operations in the
      information environment. It highlights the essential contributors to SDA
      support for mission assurance, including clearly established roles,
      intelligence preparation, accurate data, and timely information flow.

21   The page discusses the importance of Space Domain Awareness (SDA) in
      protecting space capabilities. It mentions the processing, review, and
      verification of sensor data and reports to identify potential threats and
      ensure continued availability of space capabilities. It also highlights the
      role of SDA in space battle management and the need for understanding
      adversary actions and conditions in other domains.

22   Space Doctrine Publication 3-100 discusses the importance of Space
      Domain Awareness (SDA) in space operations, including analyzing
      adversary actions and disrupting their mission planning process. It also
      highlights the role of operations in the information environment to
      influence and exploit adversary decision making.

23   The Space Force aims to promote a safe space environment through Space
      Domain Awareness (SDA). This involves sharing SDA data and providing
      SDA services to increase cooperation and collaboration. Space Situational
      Awareness (SSA) is a subset of SDA and involves collecting data on space
      objects using sensors to develop actionable information for threat warnings
      and assessment. Different types of sensors, including radar, are used to
      track and characterize resident space objects. The Space Surveillance
      Network consists of dedicated, collateral, and contributing sensors

24   This page discusses the use of radar in imaging space objects and the
      factors that affect radar's ability to detect and track these objects. It also
      describes the three types of radars commonly used for space surveillance:
      continuous wave, dish, and phased array.

25   Large-phased array radars are highly agile and flexible, making them
      effective for tracking space objects. Ground-based radars are the most
      common for space surveillance, but maritime radars on ships can also
      track some space objects.

26   This page discusses the different types of radars and sensors used for
      tracking space objects, highlighting their capabilities and limitations. It

emphasizes the cost and logistical challenges of implementing a space-based radar system. Optical and infrared sensors are also mentioned, noting their ability to provide angular accuracy but limited range determination.

27   This page provides information on the types of optical sensors used in telescopes (refracting, reflecting, and catadioptric) as well as the two modes of operation for observing space objects (sidereal track and rate-track).

28   This page discusses the operating considerations for optical and infrared sensors in the land and space domains. Ground-based sensors are limited by lighting conditions and weather, but are effective for detecting and tracking objects at high altitudes. Space-based sensors offer advantages such as look-angle diversity and the ability to track smaller objects, but are subject to space weather.

29   The page discusses different types of sensors used for space domain awareness, including optical sensors, infrared sensors, passive radio frequency sensors, and space environmental monitoring sensors. Each type of sensor has its advantages and limitations in detecting and tracking objects in space.

30   Space Doctrine Publication 3-100 discusses the importance of space domain awareness and the various sources of data and intelligence that contribute to it. It also highlights the shift from a tasked-track methodology to a search-based approach for optimal sensor employment. Factors that may result in missed collection opportunities are also mentioned.

31   The page discusses reasons for missed passes on suspicious space objects and the need for immediate action. It also highlights the importance of understanding the link and terrestrial segments of space systems for mission accomplishment and provides ways to gain awareness of these segments.

32   Space Doctrine Publication 3-100 discusses the challenges and considerations of Space Domain Awareness (SDA) data. It emphasizes the need for collecting, synthesizing, and understanding large volumes of diverse data from various sources to achieve maximum domain awareness. The accuracy, timeliness, and prioritization of SDA information are also crucial for effective operations. Additionally, an accurate catalog of objects is essential for identifying potential hazards or threats to spacecraft.

33   This page discusses the importance of environmental monitoring, conjunction assessment, and deorbit and reentry support in the context of space domain awareness. It highlights the need to identify hazards and threats to space capabilities, protect high-value assets, and provide warning and mitigation actions to avoid collisions and impacts on Earth.

34   The Space Force's Space Operations Command (SpOC) collaborates with various organizations to analyze different forms of information and develop reliable Space Domain Awareness (SDA). SpOC operates terrestrial and on-orbit assets to collect space observation data, supports SDA operations, defends US space capabilities, provides warning of harmful activity, tracks objects in Earth orbit, and maintains global awareness of space forces.

cislunar space. The requirement to develop SDA capabilities in cislunar space will continue to grow

46   Military space forces have three cornerstone responsibilities: preserving freedom of action, enabling joint lethality and effectiveness, and providing independent options. The United States Space Force executes five core competencies, including space security, combat power projection, and space mobility and logistics.

47   This page discusses various aspects of space warfare, including information mobility, space domain awareness, orbital warfare, space electromagnetic warfare, space battle management, space access and sustainment, military intelligence, and engineering and acquisition. These disciplines are essential for the United States Space Force in developing personnel and defending the space domain.

48   The page discusses the importance of cyber operations in defending and securing critical space networks and systems, as well as the need for future offensive capabilities in this domain.

NOTABLE PASSAGES

1   "Space Domain Awareness (SDA) is the requisite foundation for space operations. It encompasses the knowledge and understanding of the space environment, including the physical and virtual assets that reside within it, and the ability to characterize, track, and identify those assets. SDA provides the necessary information to support decision-making processes, enabling the effective and efficient use of space capabilities. It includes the collection, processing, analysis, and dissemination of space-related data and information, as well as the integration of that information into operational planning and execution. SDA is critical for situational awareness, space traffic management, space control, and space defense. It enables the identification and assessment of potential threats, vulnerabilities, and opportunities in the space domain, and supports the development and implementation of strategies, policies, and procedures to protect and enhance U.S. national security interests in space."

2   "Strength and security in space provides national leaders with independent options and enables freedom of action in space and other warfighting domains while contributing to international security and stability. Effective SDA is foundational for space forces to conduct prompt and sustained operations that fulfill the cornerstone responsibilities of the Space Force: preserving freedom of action in the space domain, enabling joint lethality and effectiveness, and providing independent options capable of achieving national objectives. Space Force commanders and their staffs rely on timely and actionable SDA to satisfy these responsibilities."

3   "Figure 3. History of tracked objects in Earth orbit"

4   "Space Force doctrine guides the proper use of military spacepower in support of the Service's cornerstone responsibilities. It establishes a common framework for employing Guardians as part of a broader joint force. Doctrine provides fundamental principles and authoritative guidance for the employment of military spacepower and an informed starting point for decision making and strategy development."

5   "Some of the spacecraft on orbit now possess capabilities and sensors that could threaten US, allied, or partner operations in space or terrestrially. It is now feasible for proliferated constellations comprised of thousands of small spacecraft to provide persistent, global coverage across a variety of mission sets."

6   "Combined, these factors make it imperative for the safety of space operations that the United States not only knows where objects and spacecraft are at any given time, but also how they got there, who owns them, their potential capabilities, and their operator's intent."

7   "To meet the needs of safe operations in the space domain, space situational awareness or SSA, which traditionally focused on finding, tracking, identifying, and maintaining custody of space objects as an operational task, evolved into SDA. SDA fuses intelligence about each spacecraft, their components in other domains, and the environments they operate in, to produce the most complete possible understanding of space events, threats, activities, conditions, and system components operating in the orbital, terrestrial, and link segments. SDA is the timely, relevant, and actionable understanding of the operational environment that allows military forces to plan, integrate, execute, and assess space operations."

8   "The space environment is influenced by solar activity that ebbs and flows on an 11-year cycle. Geomagnetic storms result from variations in the solar wind that produce major changes in the currents, plasmas, and fields in Earth's magnetosphere. High-speed solar wind conditions create geomagnetic storms that may last for hours or days. The largest storms that result from these conditions are associated with solar coronal mass ejections where approximately a billion tons of plasma from the sun, with its embedded magnetic field, arrives at Earth. Storms result in intense currents in the magnetosphere, changes in the radiation belts, and changes in the ionosphere, including heating the ionosphere and upper atmosphere region called the thermosphere."

9   "In 2013, a 20-meter asteroid entered Earth's atmosphere, resulting in a superbolide over Russia that damaged more than 7,200 buildings and injured nearly 1,500 people. Should a similar event occur today, the geopolitical repercussions could be catastrophic. Detecting these objects in advance of a superbolide event will be critical for de-escalation during conflict or crises. In extreme cases, asteroids and comets can pose a direct impact risk to the Earth. Such an event could have global impacts."

10   "In an increasingly contested space domain, superior knowledge of the natural environment provides space actors with the means to plan and execute operations better than their competitors and adversaries. Our adversaries are expanding their ability to understand the current and future state of the natural environment and advancing their ability to exploit that understanding to their advantage. For the Space Force,

superior knowledge of the natural environment will enable Guardians to protect and defend assigned space mission systems and to operate as an integral part of the joint and combined force despite any hazards. The United States must continuously advance its knowledge of the natural environments to increase the capability and capacity of space systems supporting freedom of access and action in the space domain without prohibitive interference from an adversary."

11   "Adversaries are aware of the military advantages the United States accrues from space capabilities, and our corresponding reliance on them. They understand how the United States and its allies use space to provide timely intelligence; communications; positioning, navigation, and timing; and missile warning capabilities. This incentivizes adversaries to attempt to deny those advantages by denying US freedom of action in space. Understanding the threat to US and allied space capabilities is a critical component of SDA."

12   "Adversaries can use high-powered lasers to damage or degrade sensitive spacecraft components, such as solar arrays. They can also use lasers to temporarily or permanently blind mission-critical sensors on spacecraft. High-powered microwave or electromagnetic pulse weapons can be used to disrupt a spacecraft's electronics, corrupt data stored in memory, cause processors to restart, or at higher power levels, cause permanent damage to electrical circuits and processors."

13   "On orbit, adversaries have demonstrated sophisticated technologies that could be weaponized to target the orbital or link segments of space systems. For example, the technology to rendezvous with other spacecraft to inspect or repair them is also technology that could be used to employ an orbital weapon. Other capabilities, such as directed energy weapons discussed above or ASATs, are also potential threats that would challenge the ability to detect, attribute, and respond to an attack. More advanced weapons could also employ proximity operations and robotic arms to grapple target spacecraft to disrupt their orientation or damage components. Effective SDA, to include timely and accurate positional and other information about on-orbit objects, which are difficult to characterize and monitor, is critical to defeating these threats."

14   "Because they are often highly visible, sometimes located outside of the United States, and more accessible than objects in space, the terrestrial segment of a space system can be a more inviting target for adversaries seeking to disrupt or degrade space capabilities. Any element of terrestrial space infrastructure, including antennas, fiber communication lines, control centers, ground radars, telescopes, or even associated personnel, are all potential targets."

15   "The advent of 'quick-response space launch vehicles,' able to rapidly supplement or reconstitute on-orbit capabilities during a conflict, creates an SDA challenge. Increasingly, these systems can expedite launch campaigns using transportable launchers, limiting warning timelines of potential threats."

16   "SDA plays a key role ensuring the United States can meet a variety of international obligations, including compliance with the Outer Space Treaty and other agreements. Effective SDA can also help the United

States determine if other nations are meeting their international obligations."

17   "SDA is achieved through the integration of intelligence, surveillance, reconnaissance, environmental monitoring, and friendly force tracking. SDA goes beyond space surveillance (object tracking, characterization, and establishing pattern of life), to include understanding intent, motive, and predicted actions across the terrestrial and link segments. SDA incorporates multiple intelligence and information capabilities, including terrestrial and space-based surveillance, space and terrestrial weather, and information and data-sharing capabilities. Operators, users, and decision makers obtain SDA through a timely appreciation of factors and actors providing awareness and object intelligence in the physical space domain, awareness of the operations and threats in the electromagnetic spectrum, awareness of systems across all domains, and awareness of dependencies on operations in other domains."

18   "SDA integrates intelligence, surveillance, and reconnaissance; environmental monitoring; and friendly and adversarial forces' information in each operational environment and across all domains. SDA also necessitates integration of information, intelligence, and sharing with allies and partners to maintain awareness of joint force dependencies on space capabilities and adversary behavior as it pertains to space operations."

19   "In any operational environment, valuable mission capabilities are likely to be threatened or attacked. SDA is critical to space operations because it allows decision makers to understand actions and events in context. This allows commanders to direct efforts more effectively and assign resources to achieve desired results, whether in support of an operation to defend an asset or to act against an adversary. Degradation or loss of SDA capabilities could undermine a commander's understanding of the operational environment, increasing the likelihood of adversary gaining an operational advantage. Attacks and natural hazards can have similar effects. It is part of SDA to understand the source and nature of anything affecting an SDA asset."

20   "Synergy throughout the operational environment is essential to effective joint operations. Space capabilities, integrated into joint planning before operations begin, support unified action and synergy within the operational environment. The joint force relies on space capabilities such as SDA to understand the operational environment. SDA contributes to mission assurance, threat warning and assessment, space battle management, and operations in the information environment, enhancing the joint force commander's understanding of the operational environment."

21   "Uncorrelated sensor observations that match a potential threat or cannot be correlated to known objects should drive C2 entities to generate analyst object element sets for follow-up tracking and identification of the object in question. Analyst objects refer to on-orbit objects tracked by the Space Surveillance Network (SSN), but with insufficient fidelity for publication in the public satellite catalog."

22   "Delegating authorities to the lowest possible level, while maintaining centralized C2, enables agility, timely delivery of SDA information, and the increased speed of execution required to support space battle management inside the adversary's decision cycle."

23   "The Space Force is committed to promoting a safe, stable, sustainable, and secure space environment through SDA information sharing. As more nations, non-state actors, commercial entities, and non-governmental organizations field space capabilities and benefit from the use of space systems, it is in our collective interest to act responsibly and enhance overall spaceflight safety."

24   "A radar's ability to detect and track an object is affected by the radar's transmitting power, the transmitter's gain, the receiver's gain, the object's radar cross-section, and the range to the object. The object's radar cross-section is how big the object appears to the radar and is a function of not only the object's physical size but how much of the radar signal reflects to the radar receiver. The radar's signal spreads as it travels to the target and then the reflection spreads out again on its way back to the radar receiver."

25   "Most large-phased array radars can form keyed fences—regions of space observed for the presence of resident space objects for which no position information exists—for un-cued detection and simultaneously use separate beams to continue tracking new objects. This makes large-phased array radars excellent choices for gaining custody of objects whose position or even existence is uncertain."

26   "The telescopes used for tracking space objects tend to have a wide field-of-view compared to radars, and report on every space object in their field-of-view, which makes them excellent instruments for searching wide areas of sky. The un-cued sensitivity of optical sensors generally varies with the diameter of the telescope's aperture or primary mirror. The larger the telescope, the harder it is to move quickly, making this a design and operational trade-off. The photometric and spectral data from optical sensors can be used to characterize an object's size, material composition, patterns of behavior, and even age. One limitation of optical sensors is that they only report angular data without range data, unless they are equipped with a rangefinder."

27   "Rate Tracking. The rate-tracking mode moves the telescope to match the motion of the object being observed so the object appears as a single point of light and the background stars appear as streaks. Since all the light from the observed object remains concentrated on one point, this method makes it possible for a telescope to pick up dimmer objects than it might see using sidereal tracking. However, it requires knowing or at least suspecting the object's motion to avoid smearing its light over the focal plane. In addition, the observed object may be hidden or obscured by a star streak passing through it. Rate tracking with a high refresh rate (very small frame or integration time) is a preferred collection method for characterizing an object."

28   "Employing space-based optical sensors in the right orbit gives the advantage of look-angle diversity that is unobtainable when using only ground sites. Additionally, due to significantly reduced range to their

collection targets, space-based optical sensors can track smaller or dimmer objects than ground-based optical sensors with equivalent apertures. Orbital assets are not disrupted by weather, time of day and atmospheric distortion that limit ground-based systems."

29 "Because infrared sensors do not require sunlight for target illumination, they can detect objects under conditions that would otherwise be non-ideal for optical sensors. Additionally, infrared sensors can detect infrared light sources that are particularly dim compared to optical light sources. These advantages create more opportunities to detect, characterize, and track objects that are distant, dim, or otherwise difficult to observe."

30 "The Space Force is moving to a search-based, task-as-required methodology that establishes a base of regular periodic searches to prevent, or at least minimize, surprise by hostile actors. The regular searches are supplemented by taskings to track specific known spacecraft or objects necessary to support missions requiring more accurate element sets or more persistent monitoring of non-cooperative spacecraft (i.e., spacecraft whose owner or operator does not provide notice of upcoming launches or maneuvers)."

31 "Missed passes on suspicious space objects should prompt immediate action to search for them (missed pass report / no-show report to the C2 agency, and immediate voice tasking of the next several sensors in view) and possibly issue warnings if it was a high probability pass. These types of actions could result in wasted time and resources if it was a low probability pass."

32 "Guardians should be capable of fusing data and information from all available and applicable sources to provide timely and actionable SDA."

33 "Conjunction assessment is the process for determining the point and time of closest approach of two tracked orbiting objects. This includes matching active spacecraft against the spacecraft catalog several times per day to identify and predict close approaches. This information is essential for protection of high-value assets or identification of evolving threats. If a close approach meets emergency reportable criteria, the spacecraft's owner/operator should be notified so they can determine the appropriate action to avoid a collision."

34 "These different forms of information analyzed together to broaden the scope of knowledge from being reactive and maintaining a catalog of objects on orbit, to predictive SDA that considers missions, intentions, and system capabilities and equips decision makers with the near real-time information necessary to act within an adversary's decision cycle."

35 "Data sharing is key to space security because it improves SDA for all parties. Effective sharing requires development of key partnerships among the DoD, the Intelligence Community, intragovernmental agencies, international bodies, civil agencies, allies, partners, academic, and commercial entities. This can even include competitors and adversaries in the appropriate circumstances where potential areas of shared interest exist such as space debris mitigation or collision avoidance. Additionally, data sharing agreements enable the US to operate from geographic locations necessary for achieving robust SDA. Effective data sharing is dependent upon timely and secure communication channels and

encryption protocols to protect sensitive information while enabling rapid decision making and response to potential threats. In addition to data sharing, collaboration on analytical processes and methods helps promote greater interoperability with partners and allies, many of whom have already integrated with US combat capability in other domains. This requires clear data sharing agreements that protect intellectual."

36   "Strengthening relationships with allies and partners, by sharing intelligence data and analytical processes and methods, is critical to promoting space security. The Space Force continually reevaluates data sharing and collaboration agreements, and security paradigms that often limit interoperability with allies and partners in space."

37   "Space Doctrine Publication 3-100, Space Domain Awareness November 2023"

38   "Electromagnetic warfare — Military action involving the use of electromagnetic and directed energy to control the electromagnetic spectrum or to attack the enemy. Also called EW."

39   "Space battle management – 1. Knowledge of how to orient to the space domain and skill in making decisions to preserve mission, deny adversary access, and ultimately ensure mission accomplishment. 2. Ability to identify hostile actions and entities, conduct combat identification, target, and direct action in response to an evolving threat environment."

40   "Space superiority — The degree of control in the space domain of one force over another that permits freedom of access and action without prohibitive interference."

41   "Challenges to security in space: space reliance in an era of competition and expansion."

42   "The Strategy is rooted in our national interests: to protect the security of the American people, to expand economic opportunity, and to realize and defend the democratic values at the heart of the American way of life."

43   "Department of Defense Electromagnetic Spectrum Superiority Strategy, October 2020 –

Addresses how DoD will: develop superior electromagnetic spectrum capabilities; evolve to

an agile, fully integrated electromagnetic spectrum infrastructure; pursue total force

electromagnetic spectrum readiness; secure enduring partnerships for electromagnetic

spectrum advantage; and establish effective electromagnetic spectrum governance to support

strategic and operational objectives. Investment in these areas will speed decision-quality

information to the warfighter, establish effective electromagnetic battle management, enable

electromagnetic spectrum sharing with commercial partners, advance electromagnetic

spectrum warfighting capabilities, and ensure our forces maintain electromagnetic spectrum

superiority."

44 "National Preparedness Strategy & Action Plan for Near-Earth Object Hazards and Planetary Defense – Guides DoD and other US government agencies and departments in developing capabilities and technologies necessary to enhance hazardous object detection and mitigation."

45 "As the likelihood of threats emanating from the cislunar regime beyond GEO transitions from part of a far-off future to an issue we expect to contend with in the foreseeable future, the urgency to obtain and maintain SDA in that regime increases."

46 "Unfettered access to and freedom to operate in space is a vital national interest; it is the ability to accomplish all four components of national power – diplomatic, information, military, and economic – of a nation's implicit or explicit space strategy. Military space forces fundamentally exist to protect, defend, and preserve this freedom of action."

47 "Orbital warfare. Knowledge of orbital maneuver as well as offensive and defensive fires to preserve freedom of access to the domain. Skill to ensure United States and coalition space forces can continue to provide capability to the joint force while denying that same advantage to the adversary."

48 "Knowledge to defend the global networks upon which military spacepower is vitally dependent. Ability to employ cyber security and cyber defense of critical space networks and systems. Skill to employ future offensive capabilities."

Image Placeholder

The AI-generated prompt was: Create a black and white illustration that captures the intense focus and determination of space operators as they monitor the vastness of outer space. The drawing should convey a sense of urgency and vigilance, with individuals diligently analyzing data and monitoring screens filled with orbiting objects. Incorporate elements that represent collaboration and data sharing, such as multiple operators working together or imagery representing the exchange of information between countries. The illustration should evoke a mood of professionalism and responsibility, emphasizing the commitment to maintaining a safe and secure space environment for military operations.

Space Doctrine Publication 3-100

SPACE DOMAIN AWARENESS

DOCTRINE FOR SPACE FORCES

UNITED STATES
SPACE FORCE

Space Doctrine Publication (SDP) 3-100, *Space Domain Awareness*
Space Training and Readiness Command (STARCOM)
OPR: STARCOM Delta 10
November 2023

Foreword

United States Space Force (USSF) doctrine guides the proper use of military spacepower in support of the Service's cornerstone responsibilities. It establishes a common frame of reference on the best way to plan and employ Space Force forces as part of a broader joint force. This doctrine provides official advice to execute and leverage spacepower using its core competencies. It is not directive—rather, it provides Guardians an informed starting point for decision making and strategy development.

Space Doctrine Publication 3-100, *Space Domain Awareness (SDA)*, aligns with current Space Force doctrine and the Chief of Space Operations' Planning Guidance. It articulates best practices and lessons learned for achieving SDA, using existing capabilities, to ensure the ability to operate safely while protecting and defending our freedom to operate in space.

Strength and security in space provides national leaders with independent options and enables freedom of action in space and other warfighting domains while contributing to international security and stability. Effective SDA is foundational for space forces to conduct prompt and sustained operations that fulfill the cornerstone responsibilities of the Space Force: preserving freedom of action in the space domain, enabling joint lethality and effectiveness, and providing independent options capable of achieving national objectives. Space Force commanders and their staffs rely on timely and actionable SDA to satisfy these responsibilities.

I encourage all Guardians to study and learn from the knowledge compiled in this publication. Semper Supra!

TIMOTHY A. SEJBA
Brigadier General, USSF
Commander, Space Training and Readiness Command

Table of Contents

Figures

Space Force Doctrine

Space Force doctrine guides the proper use of military spacepower in support of the Service's cornerstone responsibilities. It establishes a common framework for employing Guardians as part of a broader joint force. Doctrine provides fundamental principles and authoritative guidance for the employment of military spacepower and an informed starting point for decision making and strategy development. Since it is impossible to predict the timing, location, and conditions of the next fight, commanders should be flexible in the implementation of this guidance as circumstances or mission dictate.

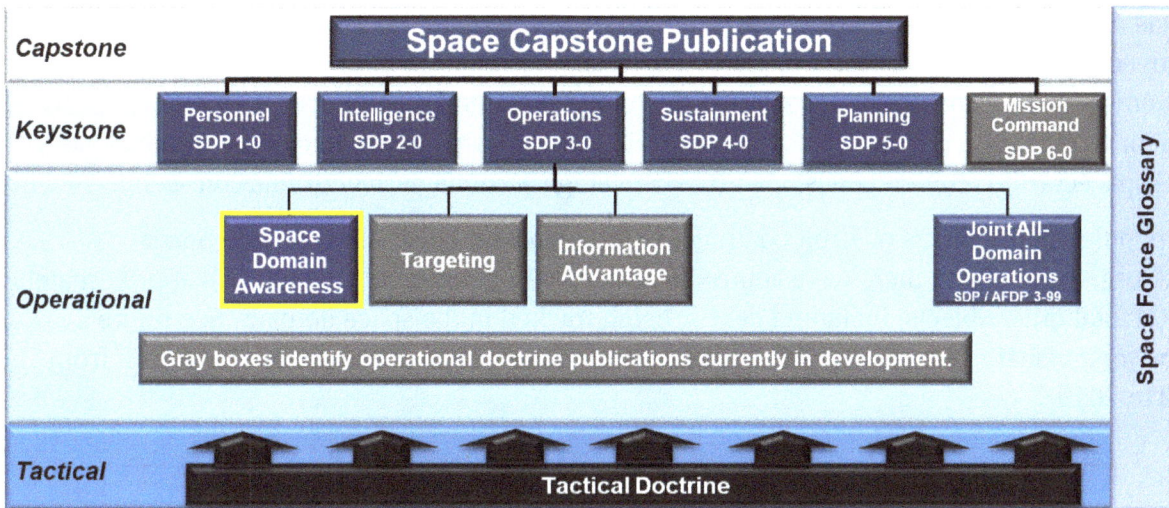

Figure 1. Space Force doctrine hierarchy

Space Doctrine Publication (SDP) 3-100

SDP 3-100, *Space Domain Awareness (SDA),* an operational-level publication, presents the Space Force's approach to establishing and maintaining SDA as part of unified action to support the freedom to operate in, from, and to space.

- Chapter 1 explains the importance of SDA and describes the space environment, to include the natural operating environment, space debris, threats, adversary use of space, and commercial space.

- Chapter 2 introduces the SDA discipline and discusses SDA in the joint operational environment.

- Chapter 3 explains SDA capabilities and discusses the orbital, terrestrial, and link segments of space systems and the data sharing environment.

- Chapter 4 discusses the roles and responsibilities of organizations that conduct SDA, contributing sensors, and data sharing.

Chapter 1: The Strategic Imperative

SDP 3-100 provides operational doctrine to the Space Force for SDA. This publication addresses the relationships between the elements of SDA and describes best practices based on existing capabilities.

Joint Publication 3-14, *Joint Space Operations*, defines the space domain as the area above the altitude where atmospheric effects on airborne objects become negligible. The United States, foreign governments, commercial industry, and academia are rapidly increasing the number of objects in the space domain. For example, in 2020 the International Telecommunication Union approved 42,000 spacecraft for a single commercial constellation. Some of the spacecraft on orbit now possess capabilities and sensors that could threaten US, allied, or partner operations in space or terrestrially. It is now feasible for proliferated constellations comprised of thousands of small spacecraft to provide persistent, global coverage across a variety of mission sets.

The population of objects on orbit continues to rise exponentially. According to Space-Track.org, in July 2023 there were approximately 8,400 operational payloads and approximately 44,900 total other objects, including debris, being tracked in the space domain. See figure 2 below for a visual representation of the increase in tracked objects and space congestion from 1960 to 2019.[2]

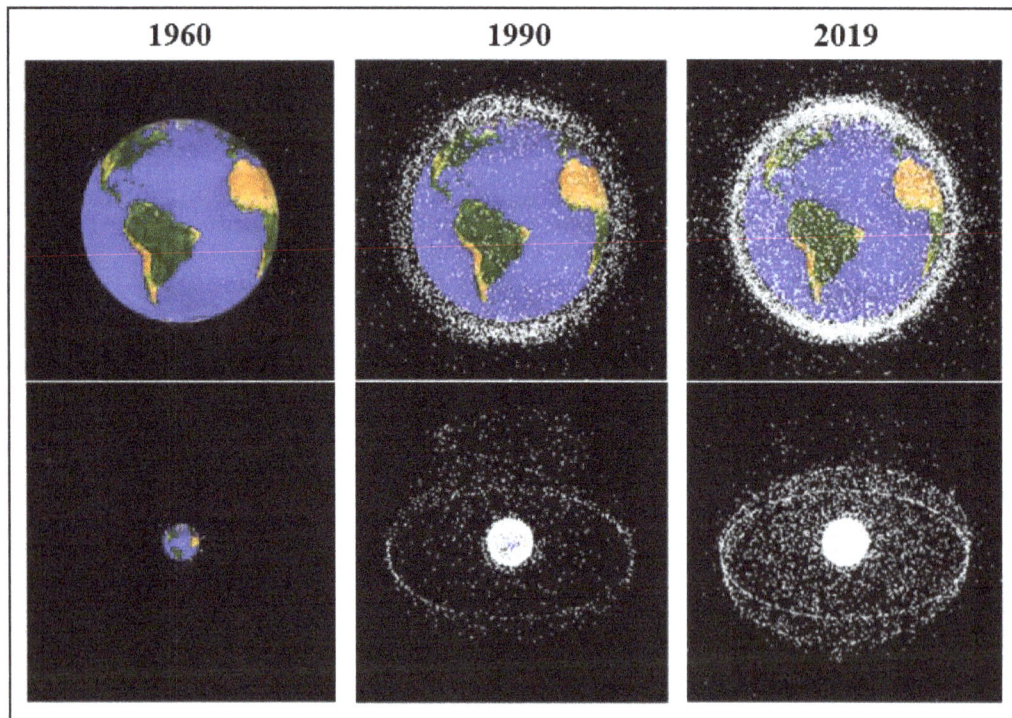

Figure 2. Visualization of orbital objects—1960, 1990, 2019
(Credit: National Aeronautics and Space Administration (NASA) Orbital Debris Program Office)[2]

The steep increase in the number of tracked items reflected in figure 3 is attributable to new state, non-state, and commercial spacecraft; decreased barriers to entry to space; large constellations of hundreds to thousands of spacecraft placed on orbit by commercial companies (e.g., Starlink); a few significant debris-creating events such as the 2007 Chinese antisatellite (ASAT) test, the 2009 Iridium-Cosmos collision, and the 2021 Russian ASAT test, and improved monitoring capabilities. At the same time, the hazards posed by the space environment and natural debris continue to threaten spacecraft and create the potential for additional debris-creating events. Combined, these factors make it imperative for the safety of space operations that the United States not only knows where objects and spacecraft are at any given time, but also how they got there, who owns them, their potential capabilities, and their operator's intent.

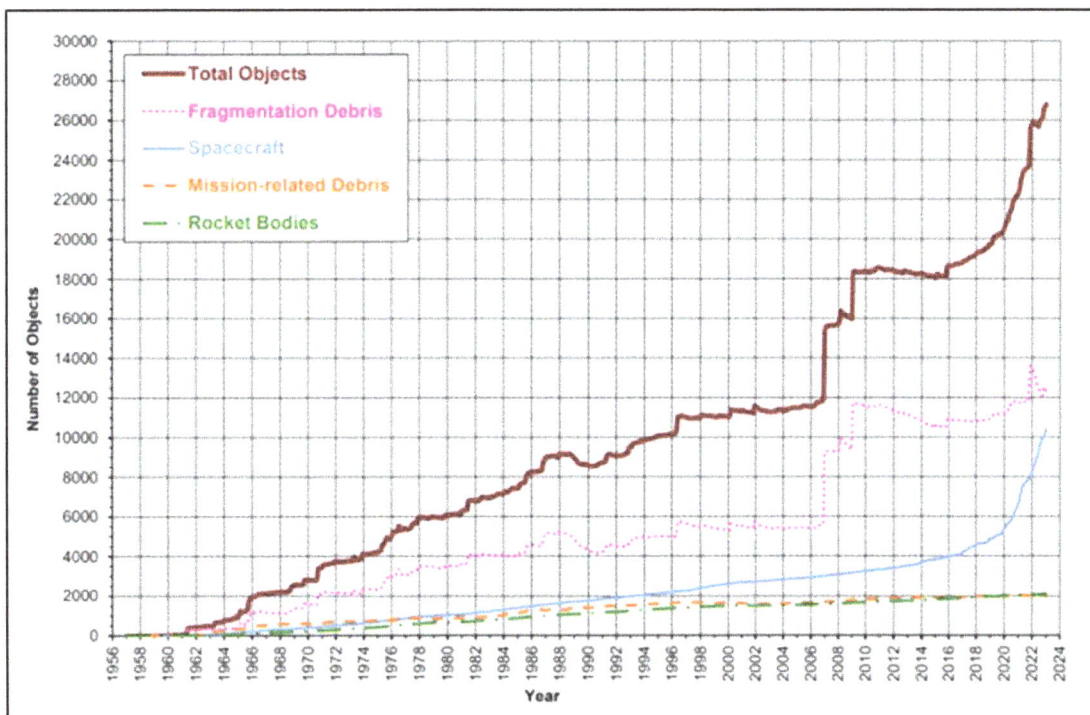

Figure 3. History of tracked objects in Earth orbit
(Credit: NASA Orbital Debris Program Office Quarterly News)[2]

At the same time, the electromagnetic spectrum and cyberspace domain are also becoming increasingly congested and competitive in the context of space operations. The number of terrestrial capabilities, nodes, and assets operating as part of the international space architecture is also rapidly increasing. As competitors continue to partner with new stakeholders, our ability to maintain the requisite awareness of these activities, the associated nodes and assets, and their use of the electromagnetic spectrum and cyberspace is vital to our understanding of the space domain and the safety of space operations.

Operating safely in the space domain requires the ability to detect, track, characterize, discriminate between, and maintain custody of increasingly smaller spacecraft and debris with increased accuracy. It requires the ability to detect, track, characterize, discriminate, and maintain custody of objects that are dimmer, more distant, or otherwise difficult to observe despite the increased proliferation of objects that are nearer and brighter. It requires the ability to rapidly identify and respond to threats and hazards, including objects that exhibit abnormal observables and patterns of life and cannot by correlated to any owner or point of origin. It also requires the ability to collect, transmit, store, and analyze an ever-increasing amount of observational data on tactically relevant timelines.

To meet the needs of safe operations in the space domain, space situational awareness or SSA, which traditionally focused on finding, tracking, identifying, and maintaining custody of space objects as an operational task, evolved into SDA. SDA fuses intelligence about each spacecraft, their components in other domains, and the environments they operate in, to produce the most complete possible understanding of space events, threats, activities, conditions, and system components operating in the orbital, terrestrial, and link segments. SDA is the timely, relevant, and actionable understanding of the operational environment that allows military forces to plan, integrate, execute, and assess space operations.

The Space Force organizes, trains, equips, and presents Guardians to deliver spacepower as part of unified action for the United States. On behalf of the joint force commander, Guardians conduct a range of operations including SDA, one of the Space Force's five core competencies (see appendix f). SDA, which includes understanding of the space environment and applies space battle management principles to provide military options and direction to Guardians, is foundational to executing all space operations across the competition continuum.

The Space Operating Environment

SDA maps the space operating environment for Guardians to conduct defensive and offensive operations in response to threats, and movement operations to sustain capabilities and achieve military objectives. The term "space terrain" describes subsets of the space domain—distinguished by unique physical characteristics and the presence of resident space objects (e.g., spacecraft, debris) in them—through the lens of military utility (e.g., battlespace). These subsets can include, but are not limited to, different orbital regimes, gravity wells, celestial objects, and the space environment (e.g., radiation belts). The Guardians as part of the joint force maintain the integrity, completeness, and responsiveness of the space domain, as well as identifying, characterizing, and mitigating risks in the space terrain.

Hazards and threats to space operations complicate the ability to develop and maintain SDA across the competition continuum. Hazards are phenomena that can cause harm resulting from natural, neutral, or unintentional activities or events. Hazards include the natural space environment, orbital congestion and constellation proliferation, and the ever-increasing volume of on-orbit debris. Conversely, threats are phenomena that can cause harm resulting from intentional activities or events. Hazards and threats challenge the safety and security of space operations, including space traffic management. Hazards and threats can negatively impact SDA by complicating conjunction warning, hindering threat identification, and presenting anomalous indications that unnecessarily consume SDA resources (e.g., sensors, communication nodes, command and control [C2] centers, planners, operators) that would otherwise be supporting space and terrestrial combatant commands.

a. **Natural Space Operating Environment.** As with other operational domains, the natural environment affects the space domain and can impact space systems and the services they provide warfighters. Those effects can both positively and negatively impact military operations across the competition continuum. Joint Publication 3-59, *Meteorological and Oceanographic Operations*, defines the space environment as the region "corresponding to the space domain, where electromagnetic radiation, charged particles, and electric and magnetic fields are the dominant physical influences, and that encompasses the Earth's ionosphere and magnetosphere, interplanetary space, and the solar atmosphere." Naturally occurring space objects, which range in size from micrometeoroids to asteroids and comets spanning multiple kilometers, are also part of the space environment. These natural aspects and phenomena of the space environment can present as hazards that impact space operations.

The collection and exploitation of information regarding the state of the natural environment is a vital component of SDA. Information concerning environmental conditions is essential to distinguish between sources of spacecraft anomalies in support of anomaly resolution, recovery, and space attack assessment processes. Observations of the natural environment come from space-based and ground-based systems operated by the Department of Defense (DoD), Department of Commerce, NASA, and other civil, academic, and international partners. The fusion and exploitation of space environment information supports overall situational awareness and forecasting of ever-changing battlespace conditions.

The space environment is influenced by solar activity that ebbs and flows on an 11-year cycle. Geomagnetic storms result from variations in the solar wind that produce major changes in the currents, plasmas, and fields in Earth's magnetosphere. High speed solar wind conditions create geomagnetic storms that may last for hours or days. The largest storms that result from these conditions are associated with solar coronal mass ejections where approximately a billion tons of plasma from the sun, with its embedded magnetic field, arrives at Earth.

> **Starlink Impact Example**
>
> SpaceX launched 49 Starlink internet spacecraft on February 3, 2022, which a geomagnetic storm impacted the following day. The geomagnetic storm increased the density of the atmosphere, increasing the drag on the spacecraft. According to SpaceX, analysis showed the increased drag at the low altitudes prevented the spacecraft from leaving safe mode to begin orbit-raising maneuvers, resulting in 38 of the spacecraft reentering the Earth's atmosphere.[6]

Storms result in intense currents in the magnetosphere, changes in the radiation belts, and changes in the ionosphere, including heating the ionosphere and upper atmosphere region called the thermosphere.

The space environment is also influenced by galactic activity. Galactic cosmic rays, for example, are high-energy charged particles that enter the solar system from all directions. These charged particles are known to disrupt and degrade spacecraft components over time and can trigger spacecraft anomalies and loss of mission capability. Galactic cosmic ray intensities fluctuate in a manner inversely proportional to the strength of the solar wind over the course of the 11-year cycle. In other words, when solar activity is at a minimum, galactic cosmic rays are at their highest levels and vice versa.

Energetic charged particles bombard space systems, penetrating spacecraft shielding and depositing a charge internally within the electronics of the spacecraft. This can cause surface and internal charging, potentially impacting memory, and degrading other sensitive components.

Currents in the ionosphere add energy in the form of heat that can increase the density and distribution of the upper atmosphere, increasing drag on spacecraft in low-Earth orbit. The local heating also creates strong horizontal variations in the ionospheric density that can modify the path of radio signals and create errors in the positioning information provided by global navigation satellite systems.[4] Perturbations in the aurora can also result in receipt of anomalous information and other adverse impacts to ground radar systems.

Understanding of the space environment, including the electromagnetic spectrum operational environment, plays a key role in mitigating its effects on space systems. For example, information on ionospheric activity assists Global Positioning System operators in making appropriate corrections within the signal to mitigate positioning errors. It can also impact spacecraft that use ultra-high frequencies for ground communications, preventing or degrading signal transmission (e.g., spacecraft commanding signals).

Naturally occurring space objects influence the space environment as well as the terrestrial environment. In space, micrometeoroids can impact spacecraft causing physical damage. Small and medium-sized asteroids routinely cross between the Earth and Moon and periodically transit inward of the geosynchronous Earth orbit (GEO) belt. Asteroids and comets also enter the Earth's atmosphere. Most of these objects are small and, upon entering the atmosphere, become bright exploding meteors known as bolides that are detectable by ground-based and space-based sensors. These energetic events occasionally rival nuclear weapons in terms of explosive output. In 2013, a 20-meter asteroid entered Earth's atmosphere, resulting in a superbolide over Russia that damaged more than 7,200 buildings and injured nearly 1,500 people. Should a similar event occur today, the geopolitical repercussions could be catastrophic. Detecting these objects in advance of a superbolide event will be critical for de-escalation during conflict or crises. In extreme cases, asteroids and comets can pose a direct impact risk to the Earth. Such an event could have global impacts.

The terrestrial environment also poses a resource protection concern for ground-based space capabilities and the terrestrial segment of space-based systems. Thunderstorms and heavy rain are known to cause disruptions to satellite communications and the link segment of space systems that use X-band or higher frequencies. Cloud cover impacts electro-optical observation systems. Radars may also be disrupted by terrestrial conditions, but to a lesser degree than optical systems. Severe weather or storm systems, such as hail, hurricanes, and typhoons, can damage or destroy ground equipment and facilities supporting space operations.

In an increasingly contested space domain, superior knowledge of the natural environment provides space actors with the means to plan and execute operations better than their competitors and adversaries. Our adversaries are expanding their ability to understand the current and future state of the natural environment and advancing their ability to exploit that understanding to their advantage. For the Space Force, superior knowledge of the natural environment will enable Guardians to protect and defend assigned space mission systems and to operate as an integral part of the joint and combined force despite any hazards. The United States must continuously advance its knowledge of the natural environments to increase the capability and capacity of space systems supporting freedom of access and action in the space domain without prohibitive interference from an adversary.

b. **Debris.** Monitoring and tracking space debris is inherent to maintaining SDA. Knowing where uncontrolled objects are and how they may hinder freedom of movement throughout the orbital regime is a key consideration when planning terrestrial or space-based operations. Not only does awareness of debris affect pre-planning maneuvers or posturing of forces, but it may also provide indications of intentional intercepts for destructive purposes. Space debris is a special case that can be considered a hazard and a threat depending upon the context.

Space debris can originate from several different sources. For example, spacecraft lifecycle operations—launch process, operating in space, and spacecraft end-of-life (disposal orbits, catastrophic failure, re-entry, etc.)—generate orbital debris. As the cost of space access declines, more entities, government and commercial, are becoming space actors, driving an increase in debris. Destructive actions such as testing or employing direct-ascent ASAT weapons, air-launched ASATs, on-orbit kinetic kill vehicles, or directed energy weapons to damage or destroy spacecraft may result in debris clouds that render the orbital regime hazardous and limit freedom of movement. The United States, China, Russia, and India have all conducted

> Space Policy Directive 3 (June 2018) designated the US Department of Commerce as the lead federal agency providing basic SSA data and Space Traffic Management services to commercial space entities, which will reduce the operational burden for the Space Force. In September 2022, the DoD and Department of Commerce signed a memorandum of agreement to implement Space Policy Directive 3.

these types of debris-causing events in the past. Prior to 2009, Air Force Space Command conducted conjunction assessments for select assets. Based upon warning notifications, owner or operators could then conduct risk assessments to determine appropriate courses of action in response to closely approaching objects. Today, the Space Force conducts conjunction predictions for all trackable orbital objects and issues warnings to active maneuverable spacecraft. Spacecraft owners or operators can avoid collisions and prevent the production of debris fields based on those warnings.

Adversary Offensive Space Operations

SDA is essential to providing commanders warning of an impending attack and the information necessary to respond. SDA does this by fusing information from SSA sensors, missile warning sensors, intelligence sources, and other sources to enable appropriate responses at all levels of operations: tactical, operational, and strategic.

Adversaries are aware of the military advantages the United States accrues from space capabilities, and our corresponding reliance on them. They understand how the United States and its allies use space to provide timely intelligence; communications; positioning, navigation, and timing; and missile warning capabilities. This incentivizes adversaries to attempt to deny those advantages by denying US freedom of action in space. Understanding the threat to US and allied space capabilities is a critical component of SDA.

As described in the Defense Intelligence Agency's "Challenges to Security in Space (2022)," potential adversaries are pursuing a range of threats to each segment of the space architecture (figure 4). These threats range from reversible to nonreversible effects. Reversible effects from denial and deception and electromagnetic warfare are nondestructive and temporary, allowing the system to resume normal operations when the attack ends. Electromagnetic weapons (including directed energy), cyberspace threats, and orbital threats can cause reversible or nonreversible effects. Nonreversible effects from kinetic energy attacks on space systems, physical attacks against space-related terrestrial infrastructure, and nuclear detonation in space would result in degradation or physical destruction of a space capability.[3] Each of these threats to space operations relies on the ability of an adversary to gain and maintain SDA.

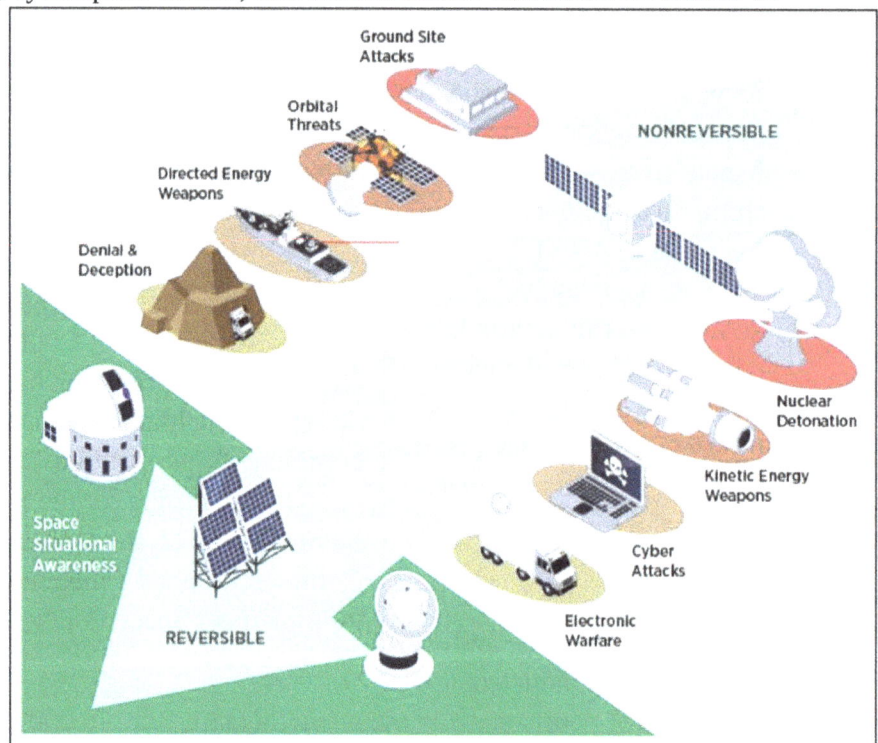

Figure 4. Threat continuum
(Credit: Defense Intelligence Agency)[3]

a. **Electromagnetic Warfare.** Space systems that rely on the radio frequency spectrum for communications or operations are subject to electromagnetic attacks targeting the link segment. Electromagnetic jamming is an attack that interferes with the link segment of space systems by generating noise in the same frequency band and within the field-of-view of the antenna on the targeted spacecraft or terrestrial receiver. Spoofing is a form of electromagnetic attack where the attacker tricks a receiver into believing a fake signal, produced by the attacker, is the real signal it is trying to receive. Adversaries can spoof the downlink from a spacecraft to inject false or corrupted data into communications systems.[5] These radio frequency signals may be critical to spacecraft telemetry, tracking, and commanding; collection; or mission data delivery.

> **Space System Segments**
>
> Space systems consist of components in three segments operating across all operational environments.
>
> The **orbital segment** includes space systems operating in the environment of the space domain.
>
> The **terrestrial segments** of space systems operate in the land, air, and maritime domains.
>
> The **link segments** of space systems operate in the information environment (cyberspace is part of the information environment) and the electromagnetic operations environment.

SDA includes threat detection and characterization in the electromagnetic spectrum. The insight gained supports offensive and defensive electromagnetic warfare operations and enables protection of the link segment. SDA systems also require use of the electromagnetic spectrum for the transmission of SDA data to users. Understanding the disposition of adversary electromagnetic warfare forces (e.g., the position and activities of known jamming apertures) is also key to attributing interference and determining an adversary's intent. Understanding of the joint force's use of a particular signal and how it supports their operations lends clarity about why an adversary may be targeting it for disruption, denial, or exploitation, and informs prioritization of restorative efforts. Verifying events for impact and intent, implementing operational procedures to manage spectrum with international bodies, and halting unintentional interference from friendly systems, contribute to the joint force's understanding, and avoid potential escalation.

The link and orbital segments of space systems can also be subject to attacks from directed energy weapons such as lasers, high-powered microwave weapons, and electromagnetic pulse weapons. Adversaries can use high-powered lasers to damage or degrade sensitive spacecraft components, such as solar arrays. They can also use lasers to temporarily or permanently blind mission-critical sensors on spacecraft. High-powered microwave or electromagnetic pulse weapons can be used to disrupt a spacecraft's electronics, corrupt data stored in memory, cause processors to restart, or at higher power levels, cause permanent damage to electrical circuits and processors.[5]

Countering the directed energy weapon threat depends on tactical understanding of where those weapons are located, either on Earth or on orbit. Intelligence should provide insight on the location of terrestrial weapons and a foundational understanding of a weapon's technical characteristics and operational limitations. SDA should also provide information on orbital weapons' locations, maneuverability, and intent. Terrestrial environmental monitoring is also of value when assessing when an adversary might have advantageous conditions for a directed energy weapon attack.

b. **Cyberspace.** Cyberspace attacks can target any of the three segments of a space system by targeting transmitted data or the terrestrial or orbital systems that use or collect the data, or that maintain the operations of the spacecraft. Cyberspace exploitation can also be used to collect intelligence by monitoring data traffic patterns (i.e., which users are communicating), monitoring the data itself, or inserting false or corrupted data into the system (e.g., spoofing). Cyberspace attacks on space systems can result in a variety of effects ranging from periodic disruptions to permanent loss of a spacecraft.[5] The nature of cyberspace makes attack attribution difficult. Because of this, cyberspace attacks, unlike other attacks, may be conducted in a peacetime environment without producing retaliatory responses.

Strong relationships between space operations, intelligence, and cyberspace professionals are essential to ensuring the ability to anticipate, or quickly attribute and recover from a cyberspace attack. Understanding if failure or degradation of a space-related service is the result of an anomaly or enemy action will influence response options.

c. **Space-Based Weapons.** On orbit, adversaries have demonstrated sophisticated technologies that could be weaponized to target the orbital or link segments of space systems. For example, the technology to rendezvous with other spacecraft to inspect or repair them is also technology that could be used to employ an orbital weapon.[3] Other capabilities, such as directed energy weapons discussed above or ASATs, are also potential threats that would challenge the ability to detect, attribute, and respond to an attack. More advanced weapons could also employ proximity operations and robotic arms to grapple target spacecraft to disrupt their orientation or damage components.

Effective SDA, to include timely and accurate positional and other information about on-orbit objects, which are difficult to characterize and monitor, is critical to defeating these threats. This challenge is amplified by the fact that an orbital engagement system can remain dormant or perform a peaceful civilian mission for days or even years before being activated.[5] Some attacks on satellites are difficult to attribute due to the method of attack or the classification of the method. In addition to SSA, information about signal activity in the electromagnetic operational environment and information about terrestrial nodes (e.g., telemetry, tracking, and commanding sites) combined with intelligence inputs can amplify our understanding of a threat and the best way to counter or defeat it.

d. **Direct-Ascent ASAT Missiles.** The orbital segment of a space system is subject to kinetic attacks by direct-ascent ASATs. Multiple countries have demonstrated this capability by testing them against their own spacecraft. Understanding known and suspected ASAT threats, to include their capabilities, employment methods, patterns of behavior, and adversary intent (to include likely targets) is critical to providing warning and enabling an effective response. Missile warning and actionable intelligence are critical to providing that warning. Missiles' relatively short flight time when targeting a spacecraft in low Earth orbit (LEO), combined with the potential ability to launch them from the ground, ships at sea, or aircraft, complicates the ability to provide timely warning intelligence. This can necessitate rapid decision making on limited information.

e. **Terrestrial Attacks.** Because they are often highly visible, sometimes located outside of the United States, and more accessible than objects in space, the terrestrial segment of a space system can be a more inviting target for adversaries seeking to disrupt or degrade space capabilities.[5] Any element of terrestrial space infrastructure, including antennas, fiber communication lines, control centers, ground radars, telescopes, or even associated personnel, are all potential targets. Ground stations can be subject to physical attacks by a variety of conventional military weapons, from guided missiles and rockets at longer ranges to small arms fire at shorter ranges. These attacks may be conducted covertly and leverage insider threats.

Mitigating physical attacks requires a self-awareness of internal dependencies and weaknesses for space infrastructure. Conducting a friendly center of gravity analysis—or, if possible, obtaining adversary center of gravity analysis—can assist in identifying critical infrastructure, information, and vulnerabilities, which may inform an assessment of most likely and most dangerous adversary courses of action. Additionally, an awareness of the disposition of adversary conventional and unconventional forces, to include missile warning and defense, can inform protection and continuity of operations efforts.

f. **High-Altitude Nuclear Detonation.** Countries possessing nuclear weapons and long-range missiles have the inherent capability to detonate a nuclear warhead in space, potentially affecting all three segments of a space system. Such an attack could have wide-reaching effects on all nations' abilities to operate in the domain. Countries less reliant on space support for their military applications may have an advantage over the United States following a nuclear detonation in space. Nuclear detonation detection sensors, missile warning systems, and reliable intelligence will be key contributors to achieving SDA capable of minimizing the effects of a nuclear attack in space. In the aftermath of such an attack, environmental monitoring capabilities are essential to understanding long-term effects on the sustainability of the space environment, including systems in all three segments.

Adversary Military Use of Space

As adversaries integrate space capabilities into their kill chains for advanced conventional weapons, it is imperative that the US maintain awareness and understanding of those systems, where they are located, and their operations. SDA enhances the joint force's ability to defend itself against those adversaries' kill chains.

a. **Intelligence, Surveillance, and Reconnaissance.** Adversary militaries use space-based intelligence, surveillance, and reconnaissance, including space object surveillance and identification systems, and missile warning systems to track, identify, characterize, and monitor US and allied forces.

b. **Communications.** Spacecraft communications are a means for global command and control used by adversary militaries.

c. **Positioning, Navigation, and Timing**. Foreign positioning, navigation, and timing providers give adversaries an independent ability to navigate and provide guidance to precision weapons.

d. **Quick-Response Launch.** The advent of "quick-response space launch vehicles," able to rapidly supplement or reconstitute on-orbit capabilities during a conflict, creates an SDA challenge. Increasingly, these systems can expedite launch campaigns using transportable launchers, limiting warning timelines of potential threats.

Commercial Space

As the number of global space actors continues to increase, achieving and maintaining SDA of space activities and space systems operated by that broader set of countries, consortia, and corporations is essential for decision making on issues such as: effectively leveraging commercial space services, understanding adversary use of third-party space capabilities, ensuring safety of flight for US national security space missions, and meeting US obligations under international treaties.

a. **Integrating Commercial Space Services.** As the private sector invests more heavily in space capabilities, integrating those innovations with Space Force capabilities can enhance operational effectiveness, resilience, and deterrence. This may inform commanders' apportionment of SDA assets to prioritize Space Force monitoring for targets where there is little or no commercial collection capability available or to re-prioritize DoD assets if there is sufficient commercial capability to achieve the desired effect.

b. **Understanding Adversary Use of Commercial Space.** Commercial space firms broaden the availability of space services, allowing adversaries to purchase or negotiate access to communications, intelligence, surveillance, reconnaissance, and launch capabilities. Leveraging SDA, intelligence, and other sources of information about which commercial entities offer space capabilities to our adversaries, can enable whole-of-government options for the United States to limit adversary access to militarily valuable services.

c. **Ensuring Safety of Flight.** Space activities power the global economy; underpin US, allied, and partner national security; and improve the daily lives of people around the world. The proliferation of constellations made up of hundreds or thousands of spacecraft necessitates constant monitoring as new spacecraft join these constellations and older systems fail or are deorbited. This complicates the ability to maneuver and increases the odds of collision. The Space Force also has an obligation to provide safety of flight support to NASA's human spaceflight missions.

d. **Meeting Treaty Obligations.** SDA plays a key role ensuring the United States can meet a variety of international obligations, including compliance with the Outer Space Treaty and other agreements. Effective SDA can also help the United States determine if other nations are meeting their international obligations.

Chapter 2: SDA Operations

Since the Soviet Union's launch of Sputnik I on 4 October 1957, and especially since the 2007 Chinese ASAT test, spacecraft and debris on orbit continue to proliferate rapidly. Simultaneously, global reliance on space-based capabilities for both civilian and military applications continue to accelerate. The increased commercialization of LEO and increases in human spaceflight, to include space tourism, have elevated the need to operate safely and securely in the domain. This evolving environment, combined with increasing adversary threat capabilities, drives the need for more accurate and timely awareness of everything occurring in space and any activities in other domains or operational environments that could negatively affect, or positively enable, space operations. The demand for better awareness to ensure the safety, stability, and security of space activities will undoubtedly continue to increase.

Introduction to Space Domain Awareness

Obtaining and maintaining domain awareness is fundamental to successful operations in any environment. Commanders must understand their operational environments to identify hazards, threats, consequences, opportunities, and risk when planning and executing operations. Awareness of the space domain is no different. SDA is achieved through the integration of intelligence, surveillance, reconnaissance,

> **Space Domain Awareness** – The timely, relevant, and actionable understanding of the operational environment that allows military forces to plan, integrate, execute, and assess space operations. (Joint Publication 3-14, *Joint Space Operations*)

environmental monitoring, and friendly force tracking. SDA goes beyond space surveillance (object tracking, characterization, and establishing pattern of life), to include understanding intent, motive, and predicted actions across the terrestrial and link segments. SDA incorporates multiple intelligence and information capabilities, including terrestrial and space-based surveillance, space and terrestrial weather, and information and data-sharing capabilities. Operators, users, and decision makers obtain SDA through a timely appreciation of factors and actors providing awareness and object intelligence in the physical space domain, awareness of the operations and threats in the electromagnetic spectrum, awareness of systems across all domains, and awareness of dependencies on operations in other domains (see figure 5).

SDA is fundamental to achieving the Space Force's cornerstone responsibilities of preserving freedom of action, enabling joint lethality and effectiveness, and providing independent options. While understanding the operational environment is a prerequisite for the joint force commander to execute operations, the vast distances, orbital constraints, and physical characteristics associated with space operations present some unique challenges. Space operations rely on capabilities provided by systems operating in all operational domains and are supported by operations in the information and electromagnetic operational environments. SDA is the Space Force's approach to understanding the operational environment across all of these. It gives the joint force insight into activity and objects throughout the operational environment that could

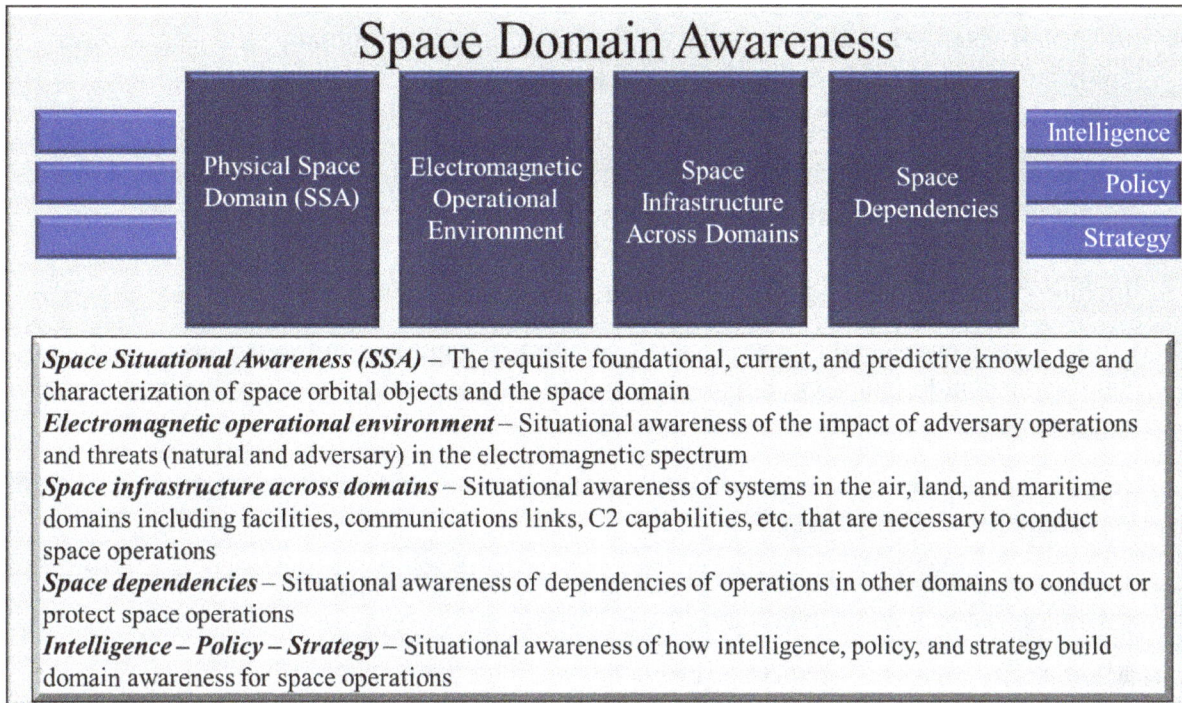

Figure 5. Space domain awareness

affect space operations. SDA integrates intelligence, surveillance, and reconnaissance; environmental monitoring; and friendly and adversarial forces' information in each operational environment and across all domains. SDA also necessitates integration of information, intelligence, and sharing with allies and partners to maintain awareness of joint force dependencies on space capabilities and adversary behavior as it pertains to space operations. SDA covers a wide breadth of elements including efforts to:

a. Understand and characterize the behavior of physical objects in, from, and to space, which includes but is not limited to operational spacecraft, meteoroids, space debris, solar wind, radiation belts, and atmospheric drag.

b. Characterize and understand capabilities and intent of adversary space systems to conduct threat assessment and develop appropriate countermeasures or response strategies.

c. Understand and monitor the effects of the natural space environment (e.g., solar weather, scintillation) their effects on space operations.

d. Understand, monitor, and characterize friendly, neutral, and adversarial use of the electromagnetic spectrum for space operations, to include uplink, downlink, and crosslink signals.

e. Detect, attribute, and characterize electromagnetic spectrum usage patterns to detect natural or manmade, unintentional or intentional, activities such as jamming, spoofing, or other forms of electromagnetic interference to maintain the integrity and functionality of systems relying on the electromagnetic spectrum.

f. Gain and maintain foundational, current, and predictive knowledge and characterization of space-related infrastructure in the air, land, and maritime domains; the information environment (to include cyberspace); and the electromagnetic spectrum operations environment—all of which enable space operations.

g. Maintain awareness of friendly, neutral, and adversarial infrastructure (e.g., hardware, equipment, and networks) required to operate or exploit a spacecraft or enable space-related operations, within or outside of the space domain—relevant information may include but is not limited to location, capability, readiness, activity, or intent.

h. Understand hazards and threats to space operations, including those to SDA collection assets that could impact a commander's ability to assess and understand the operational environment.

i. Maintain awareness of friendly, neutral, and adversarial dependencies on organic, shared, or commercial space effects.

j. Understand joint force dependencies on space capabilities to prioritize defensive space operations as well as recovery and restoral efforts.

k. Maintain awareness of adversary dependencies to identify opportunities for offensive operations and employ them at the right time and place for maximum effect.

l. Employ processes and procedures to fuse, correlate and integrate multi-source data into an operational picture and enable decision-making for operations at all levels and in all domains.

m. Employ processes and procedures required to process, exploit, and disseminate the right information to the right people at the right time.

In any operational environment, valuable mission capabilities are likely to be threatened or attacked. SDA is critical to space operations because it allows decision makers to understand actions and events in context. This allows commanders to direct efforts more effectively and assign resources to achieve desired results, whether in support of an operation to defend an asset or to act against an adversary. Degradation or loss of SDA capabilities could undermine a commander's understanding of the operational environment, increasing the likelihood of adversary gaining an operational advantage. Attacks and natural hazards can have similar effects. It is part of SDA to understand the source and nature of anything affecting an SDA asset.

Guardians deliver SDA support to joint all-domain operations by identifying, understanding, and forecasting adversary capabilities, actions, and intent. Critical products based on high-quality data, of sufficient fidelity, timeliness, and confidence, enable decision makers to execute faster decision cycles than the adversary. Leveraging the support of allies and partners provides improved access to intelligence, data products, and geography, enhancing SDA and allowing the United States to meet its worldwide mission requirements.

Understanding an adversary's SDA capability is also critical to shaping US and allied actions in all domains and operational environments. By understanding an adversary's SDA capabilities and gaps, commanders can make better-informed decisions about what to conceal and what to reveal during space operations.

SDA in the Joint Operational Environment

Synergy throughout the operational environment is essential to effective joint operations. Space capabilities, integrated into joint planning before operations begin, support unified action and synergy within the operational environment. The joint force relies on space capabilities such as SDA to understand the operational environment. SDA contributes to mission assurance, threat warning and assessment, space battle management, and operations in the information environment, enhancing the joint force commander's understanding of the operational environment.

a. **Mission Assurance.** SDA contributes necessary foundational information to support defensive, reconstitution, and resilience actions as part of mission assurance for a wide range of operations. A proactive approach to mission assurance is necessary to provide capabilities to the joint force, when needed. Spaceflight safety actions, anomaly resolution, constellation configuration, and payload calibration are examples of capabilities that derive mission assurance support from SDA. The following enabling functions are essential contributors to SDA support for mission assurance:

   1. **C2**. Clearly established supporting and supported authority roles within a mission command construct can more effectively meet the needs of the joint force and optimize mission assurance. SDA capabilities centralize data collection while employing dispersed sensors and sources in support of a mission command construct.

   2. **Intelligence Preparation of the Operational Environment/Operational Preparation of the Environment.** A strong and purposeful intelligence baseline on which to build more understanding and inform mission assurance operations is essential. Robust intelligence, combined with up-to-date blue-force status, and active environmental monitoring integrated into daily planning for space operations activities informs the understanding of the space domain necessary for mission assurance.

   3. **Accuracy and Quality of Data.** Data needs to be highly accurate, useful for targeting, and interoperable with mission and C2 systems to be actionable by the joint force for mission assurance purposes. Information systems, mission systems, and operational practices are evolving to support emerging space capabilities and defend against current and anticipated threats.

   4. **Timeliness.** To compete and succeed operationally in an agile combat environment, SDA information and derived insights need to flow rapidly to commanders. Relevant and reliable information should be available to commanders at the speed of need. This means sensors and data sources are responsive to information requests and provide information on tactically relevant timelines.

b. **Threat Warning and Assessment.** Threat warnings and assessment is the ability to assess indications and warn decision makers of potential or actual attacks, weather effects, and other risk factors affecting operations in all domains. Threat warnings and assessment processing is done at the sensor, C2, and intelligence organization levels. Guardians process, review, and verify sensor data and reports against existing threat profiles. Upon meeting certain criteria, C2 and intelligence authorities execute the decisions necessary to protect and ensure continued availability of space capabilities. When changes meet criteria for significant or critical processing, operators process them for immediate follow-up and alert the appropriate C2 centers.

C2 and intelligence organizations process, review, and verify sensor reports, track data, and uncorrelated track data against existing threat profiles. Uncorrelated tracks refer to observations that are not linked to any object already in the DoD-maintained satellite catalog of objects tracked in Earth orbit. Orbital analysts work in concert with operational intelligence analysts to identify objects and their threat potential, to provide timely, appropriate warning to command authorities and stakeholders.

Uncorrelated sensor observations that match a potential threat or cannot be correlated to known objects should drive C2 entities to generate analyst object element sets for follow-up tracking and identification of the object in question. Analyst objects refer to on-orbit objects tracked by the Space Surveillance Network (SSN), but with insufficient fidelity for publication in the public satellite catalog.

Other threat warnings and assessment may come from overhead persistent infrared data, passive radio frequency collections, electromagnetic warfare assets, satellite telemetry reports, intelligence sources, or even facility security reports (e.g., physical threats or incidents). The United States could also use treaty notifications or air and maritime closure notifications (e.g., closures of airspace that indicate rocket or missile launches or tests) to deduce threat warnings and assessment of upcoming space events and trigger specified and implied taskings. Responsibility and authority reside with the primary C2 entity, but sensor operators should integrate and coordinate operations with each other to fuse multiple sources of intelligence.

Reports of cyberspace intrusions, electromagnetic interference, and facility security incidents can be correlated with similar reports or activity in the same geographic region to find patterns of behavior indicating hostile action. Examples include cyberspace attacks from the same origin or using similar techniques, interference at the same frequencies or in the same area of responsibility, or multiple physical attacks within a particular region or against similar types of space-related facilities (e.g., Satellite Control Network sites). Within the Space Force, combat deltas work together to enable tight coordination between SDA and space intelligence units detecting, understanding, and prioritizing indications and issue warnings to affected systems and organizations.

c. **Space Battle Management.** SDA is key to successful space battle management. In addition to understanding the orbital position of resident space objects, and the state of the operational environment, Guardians require and in-depth understand of adversary actions and intent, and conditions in other domains affecting space operations, to determine whether anomalous activity is a potential threat to US, allied, or partner space operations. SDA enabled space battle management supports space superiority through understanding and analyzing an adversary's actions, expected and unexpected, to best plan for disrupting the adversary's mission planning process. Delegating authorities to the lowest possible level, while maintaining centralized C2, enables agility, timely delivery of SDA information, and the increased speed of execution required to support space battle management inside the adversary's decision cycle.

d. **Operations in the Information Environment**. Operations in the information environment allow the joint force to shape the operational environment in all domains. These operations affect drivers of behavior by informing audiences; influencing foreign actors; and attacking and exploiting relevant actor information, information networks, and information systems. Every interaction in space between military, government, civil, and commercial entities creates a pattern of behavior that communicates intent and establishes, reinforces, or diminishes norms. In turn, SDA data and information enable attribution and assessment of intent. SDA data and information can be applied, integrated, or synchronized with other means by leveraging the information environment to influence, disrupt, corrupt, or usurp the decision making of adversaries.

Chapter 3: SDA Capabilities

The Space Force is committed to promoting a safe, stable, sustainable, and secure space environment through SDA information sharing. As more nations, non-state actors, commercial entities, and non-governmental organizations field space capabilities and benefit from the use of space systems, it is in our collective interest to act responsibly and enhance overall spaceflight safety. To achieve effective SDA, the Space Force seeks to increase cooperation and collaboration with partners and space-faring entities through the exchange of SDA data and provision of SDA services.

Space Situational Awareness (SSA) Capabilities

SSA, a subset of SDA, is the requisite foundational, current, and predictive knowledge, and characterization of space objects within the space domain. That knowledge is developed through the collection of data primarily by the SSN, a combination of sensors used to detect, track, and identify resident space objects (both natural and human made).

The Space Force uses a combination of sensors, most of which are part of the SSN, to detect, track, identify, and characterize space objects in orbit to develop SSA. These systems feed data to terrestrial systems where the information is stored and processed to develop actionable information. That information is used for proper characterization of spacecraft on orbit and provides timely threat warnings and assessment to appropriate decision makers after data integration and exploitation.

Guardians use different types of sensors for tracking a variety of resident space objects. The effectiveness of the different types of sensors varies based on several factors to include the physical characteristics of the object, the object's orbit, and the capabilities of the sensor.

> **Categories of Space Surveillance Network Sensors**
>
> A <u>dedicated sensor</u> is a sensor that is operationally controlled by United States Space Command (USSPACECOM) with a primary mission of space surveillance support.
>
> A <u>collateral sensor</u> is also one that is operationally controlled by USSPACECOM and provides space surveillance support, but space surveillance support is not the primary mission. For example, sensors whose primary mission is missile warning or missile defense.
>
> <u>Contributing sensors</u> are those owned and operated by other agencies or organizations, including capabilities obtained through foreign or domestic partnerships (government, civil, academic, commercial, etc.) that provide space surveillance support upon request.

   a. **Radar**. Radars are active collectors, using a known radio frequency signal to illuminate an object and determine its position and motion. Radars can operate at any time of day and are not limited by clouds, which makes them a dependable source for tactical track applications. The algorithms developed and tuned over decades by the United States Air Force are adept at using range information to track an object's speed and path of motion. Radars can also collect special observations to characterize an object's size, shape, and orientation. For example, using range and Doppler information to produce radar images is a technique used for imaging space objects.

Most radars provide information in terms of a bearing (azimuth and elevation) to target coupled with the range and usually rate of change of the range to target (abbreviated to "range rate"). Some radars provide information in terms of the Doppler shift (change in frequency of the reflected signal) rather than directly measuring the range rate.

A radar's ability to detect and track an object is affected by the radar's transmitting power, the transmitter's gain, the receiver's gain, the object's radar cross-section, and the range to the object. The object's radar cross-section is how big the object appears to the radar and is a function of not only the object's physical size but how much of the radar signal reflects to the radar receiver. The radar's signal spreads as it travels to the target and then the reflection spreads out again on its way back to the radar receiver.

This reflection is affected by the shape of the object as well as its surface properties (some materials reflect better than others). It is also affected by the radar frequency: a shorter frequency means a longer wavelength and some of the signal may be scattered when the wavelength is longer than the object itself. The radar signal may also experience scattering or reflection if there is material between the radar and the object.

Radars and their waveforms are therefore designed for their specific missions. When tasking available space surveillance radars, operators must take the individual capabilities, as well as geographic coverage, into account. The abilities of radars to successfully detect and track an object are impacted by effects of ionospheric disturbances, terrestrial weather, and the power-to-range availability.

1. **Types of Radars**. Three types of radars are generally used for space surveillance: continuous wave, dish, and phased array.

   i.   Continuous wave radars send out a continuous wave of radar signals so that anything crossing that wave (or fence) reflects the signal and is detected. These can be the least expensive radars to build and operate, but also provide the least usable information since the single hit observation only provides a detection and position in space but cannot provide direction or magnitude of motion.

   ii. Dish or mechanical radars are relatively inexpensive to build and operate. They concentrate all their power in a single beam, which is the instantaneous field-of-view for the radar. A dish radar can only track objects within its instantaneous field-of-view and frequently, only one object at a time. This field-of-view limitation also means dish radars need cueing from another sensor to detect and track a specific object. The agility of dish radars—moving from target to target or tracking a fast-moving target—is constrained by the speed the mounts can move the dish. A dish radar pointed near the horizon also suffers from atmospheric losses.

   iii. Phased array radars use an array of dipole antennas to form their signal and steer the beam electronically rather than mechanically. The large-phased array radars used to track space objects have power rated in megawatts and enough signal processing to let them track tens to hundreds of objects simultaneously. This agility and flexibility are some of the greatest strengths of using phased arrays. Most large-phased array radars can form keyed fences—regions of space observed for the presence of resident space objects for which no position information exists—for un-cued detection and simultaneously use separate beams to continue tracking new objects. This makes large-phased array radars excellent choices for gaining custody of objects whose position or even existence is uncertain.

The boresight of a phased array extends perpendicularly from the center of the array face and the radar experiences signal losses as the signal gets further off the boresight. The field-of-regard—the area in which a beam can be steered electronically—extends to 60 degrees off boresight, which means a total field-of-regard of approximately 120 degrees for each array face. Phased arrays whose primary mission is missile warning or missile defense (i.e., collateral sensors) tend to be built with vertical or vertically inclined faces to let them establish fences at or near the horizon to ensure early detection of missile launches. Large-phased array radars designed specifically to track objects in space are oriented with a horizontal face which means it has a minimum effective elevation and cannot see objects near the horizon. Large-phased array radars are the most expensive radars to build and operate but they are also the most effective.

2. **Radar Operating Considerations**. Radars can operate from any of the physical domains (land, sea, air, space). However, not all radars are equally suited for tracking space objects.

   i. **Ground-Based Radar**. Ground-based radars constitute the preponderance of SSA sensors. Ground-based radars have large physical and logistical requirements, and interference with nearby communities must be considered. Limitations on where they can be built also affect their geographic coverage—the area of space they can observe and collect. New radars also require frequency allocations to operate in the desired locations. However, it is easier to address the limitations of power, communications, and logistics for ground-based radars than for other basing modes (e.g., sea, air, or space). Also, once built, ground-based radars tend to have greater longevity than other radars.

   ii. **Maritime Radar**. Most large ships and even small yachts employ navigation radars. Ships designed to track missiles generally have large radars with sufficient power to detect and track at least some space objects. Maritime radars have an advantage of mobility and ability to get closer to non-cooperative launching countries. However, the power available to operate the radar is limited by the space available in the ship's hull to generate power and store the necessary fuel. Communication links to transport data to C2 centers are another limitation for maritime radars. Corrosion control must also be attended to rigorously for platforms at or near the sea and the platform must be resupplied either at sea or by returning to port. The logistics of using maritime radars capable of tracking space objects suggests they should be reserved for the highest priority missions or targets of opportunity.

iii. **Airborne Radar**. Most aircraft flown by the military have radars on board, but these radars are low powered compared to radars used for space surveillance. In addition, the radars on aircraft are designed to operate against airborne or ground targets, so they do not look upwards unless the aircraft is oriented to direct the radar beams upward.

iv. **Space-Based Radar**. Space-based radar has the greatest flexibility and agility in tracking space objects; however, the cost to build, launch, operate, and replenish spacecraft has prevented the fielding of a space-based radar for tracking space objects. The effect of range on radar sensitivity means a space-based radar could detect very small objects close to it. Like all radars, the viewing capability will be limited by the power available. Angular crossing speeds are another limiting factor for using radars in space to track other space objects.

b. **Optical and Infrared Sensors**. Optical and infrared sensors are generally passive collectors, using the sun to illuminate the target object in the case of optical sensors or sensing infrared light (heat) emitted by the object in the case of infrared sensors. The sensitivity of these sensors is limited by the distance from the sensor to the target. The sensitivity varies inversely with the square of the range. The further away the target is, the larger it must be for the sensor to detect it. Other limitations include interference from bright light sources (e.g., Sun, Moon, Earth, bright foreground objects) as well as physical obstructions (e.g., weather systems, local terrain, and nearby equipment or facilities).

Optical and infrared sensors generally provide excellent angular accuracy but cannot accurately determine the range to the target. As a result, they require either longer tracks or stereoscopic tracks—viewing from two or more geographically separated locations—to determine an object's orbit with reasonable accuracy. Optical and infrared sensors with relatively wide (one square degree or more) field-of-view are also well suited for searching for unknown objects or objects with only a roughly known location.

The telescopes used for tracking space objects tend to have a wide field-of-view compared to radars, and report on every space object in their field-of-view, which makes them excellent instruments for searching wide areas of sky. The un-cued sensitivity of optical sensors generally varies with the diameter of the telescope's aperture or primary mirror. The larger the telescope, the harder it is to move quickly, making this a design and operational trade-off. The photometric and spectral data from optical sensors can be used to characterize an object's size, material composition, patterns of behavior, and even age. One limitation of optical sensors is that they only report angular data without range data, unless they are equipped with a rangefinder.

1. **Types of Optical Sensors.** There are three basic types of telescopes: refracting, reflecting, and catadioptric.

    i. **Refracting.** Refracting telescopes use lenses (and sometimes prisms) to focus light down to the focal plane sensor. They lose some efficiency as light passes through each lens. Also, the telescope becomes heavier and longer with bigger lenses. Binoculars or sighting telescopes, primarily used for distance viewing in daylight, are examples of refracting telescopes.

    ii. **Reflecting.** Reflecting telescopes use mirrors to reflect and focus light down to the focal plane sensor. They are more efficient than refracting telescopes and large mirrors are easier and more cost-effective to build than large lenses. Reflecting telescopes are commonly used for observing objects in space.

    iii. **Catadioptric.** Catadioptric telescopes use a combination of mirrors and lenses. This allows the telescope to have a large primary mirror for efficiently collecting light while the lenses can be used to correct aberrations in the light path. Most of the telescopes used for tracking space objects (e.g., the Ground-based Electro-Optical Deep Space Surveillance systems) are catadioptric designs.

2. **Modes of Operation**. There are generally two tracking modes used for operating telescopes to observe space objects: sidereal track and rate-track.

    i. **Sidereal Tracking.** Sidereal tracking moves the telescope to account for the rotation of the Earth, so the background stars appear as fixed points of light. Space objects are identified as streaks against the fixed star background. The timing for the start and end of a collection frame is known and the end points of the streak are measured resulting in observations with known times and coordinates in right ascension and declination. Sidereal track matches standard astronomy practices so it can adapt to incorporate improvements from the astronomy community. It is well-suited for un-cued detection and tracking and is the default operating mode for the Ground-based Electro-Optical Deep Space Surveillance systems.

    ii. **Rate Tracking.** The rate-tracking mode moves the telescope to match the motion of the object being observed so the object appears as a single point of light and the background stars appear as streaks. Since all the light from the observed object remains concentrated on one point, this method makes it possible for a telescope to pick up dimmer objects than it might see using sidereal tracking. However, it requires knowing or at least suspecting the object's motion to avoid smearing its light over the focal plane. In addition, the observed object may be hidden or obscured by a star streak passing through it. Rate tracking with a high refresh rate (very small frame or integration time) is a preferred collection method for characterizing an object.

3. **Optical and Infrared Sensor Operating Considerations.** Like radars, optical and infrared sensors may be deployed in many of the physical domains. However, optical, and infrared sensors currently operate only in the land and space domains.

    i. **Ground-Based Optical and Infrared Sensors.** Ground-based optical sensors are the least expensive to operate and extend globally because they do not need to be very large to be effective. These sensors are typically limited to night-time operations as they are subject to solar and lunar exclusions (sections of space where space objects cannot be observed or tracked due to lighting conditions). Susceptibility to weather and the day/night cycle makes ground-based optical sensors ill-suited for missions that may occur during daylight or in periods with extensive cloud cover. However, they play a vital role in detecting and tracking objects at altitudes above 10,000 kilometers. The Space Force also uses ground-based optical sensors with relatively wide field-of-view to perform continuous, rapid searches within 15 degrees of the equator and to collect photometric signature information. Ground-based optical sensors can track objects in LEO but typically can only see them briefly, twice per day, near sunrise and sunset when the object is still illuminated by the sun (is not eclipsed by the Earth) and is not in a solar exclusion area. To be effective, ground-based optical sensors used to track objects in LEO should have a wide field-of-view or agile mounts to accommodate the high angular motion to maximize the time they are able to view objects while they are in view. Ground-based infrared sensors, unlike optical sensors, are generally large and expensive. However, larger ground-based infrared sensors have more opportunities to track space objects because they do not rely on solar illumination. Satellite laser range trackers are a special type of ground-based optical sensors augmented with a laser to determine the range to target, offering the greatest track accuracy of any optical sensors.

    ii. **Space-Based Optical and Infrared Sensors.** The Space Force uses on-orbit payloads with optical sensors to contribute to SSA development. These sensors operate in various orbital planes and regimes providing metric observations for maintaining custody of payloads, rocket bodies, and debris to ensure safety of flight and allow time for evasive actions to avoid potential collisions. In addition, they also contribute to characterization of objects to support anomaly resolution, such as to identify a solar panel that failed to deploy properly. They can also be used to characterize adversary systems to identify threats or provide awareness of capabilities, or even lack of capability at a given time.

Employing space-based optical sensors in the right orbit gives the advantage of look-angle diversity that is unobtainable when using only ground sites. Additionally, due to significantly reduced range to their collection targets, space-based optical sensors can track smaller or dimmer objects than ground-based optical sensors with equivalent apertures. Orbital assets are not disrupted by weather, time of day and atmospheric distortion that limit ground-based systems. However, space based-optical sensors are subject to space weather which can corrupt data files or damage sensors. As with ground-based optical sensors, optical sensors in LEO are subject to solar and lunar exclusions. Optical sensors in GEO are also subject to solar and lunar exclusions, but at different times and angles than ground-based and LEO optical sensors due to their orbital characteristics. Therefore, ground-based optical sensors and optical sensors across different orbits can complement each other. The utility of space-based sensors for tactical operations can be limited by delays between collection and downlink opportunities.

Space-based infrared sensors offer additional advantages over optical sensors. Because infrared sensors do not require sunlight for target illumination, they can detect objects under conditions that would otherwise be non-ideal for optical sensors. Additionally, infrared sensors can detect infrared light sources that are particularly dim compared to optical light sources. These advantages create more opportunities to detect, characterize, and track objects that are distant, dim, or otherwise difficult to observe.

c.   Other Capabilities.

1. **Passive Radio Frequency Sensors**. Passive radio frequency sensors use the signals transmitted from spacecraft to determine their position and motion. Passive radio frequency sensors can be used to create or update element sets and vectors of target spacecraft. Unlike radar, passive radio frequency sensors do not emit a signal, but instead depend on the time difference of arrival or frequency difference of arrival of the signal. Passive radio frequency only deals with the external characteristics of the signal (frequency, waveform) and does not try to decrypt or get into the internal characteristics of the signal.

    Passive radio frequency sensors offer great persistence, tracking all day in all weather, rapid revisit, and an ability to identify and characterize the spacecraft they track. They can also help develop patterns of life and behavior which can be used to generate warning intelligence when the spacecraft deviates from its established patterns. The major limitation of passive radio frequency is this technique does not work for objects that do not emit a signal (e.g., debris) and cannot be depended on for a target that might turn off its signal.

2. **Space Environmental Monitoring Sensors.** Space-based and ground-based systems provide global observations of the sun and natural space environment. Space-based collection is primarily done by civil and international partners with sensors measuring solar activity in several emission bands (e.g., visible, X-ray, ultra-violet); magnetic field intensity; and plasma, radiation, and energetic charged particle emissions. DoD energetic charged particle sensors provide spacecraft operators awareness of the environment around their system for anomaly detection, assessment, and resolution.

    These space-based systems operate in multiple orbits (e.g., LEO and GEO) and orbital regimes (e.g., geocentric, cislunar, solar) based on the phenomena being observed. Ground-based systems within the DoD operate at locations across the globe measuring solar activity—in multiple emission bands—across the electromagnetic spectrum, as well as passive systems that monitor the ionosphere for scintillation and irregularities at ultra-high frequency and L-band frequencies. Civil and international partners provide additional ground capabilities, primarily augmenting measurements of the lower ionosphere. As of July 2023, the United States Air Force processes space environmental data on behalf of the Space Force to provide products Guardians use to conduct anomaly resolution and to the joint force for mission planning and execution. The space environmental data mission is in the process of being transferred to the Space Force.

3. **Intelligence Sources.** Intelligence Community assets contribute to SDA by providing a wide range of data and assessments, including information about planned launches, adversary space capabilities, and other activities that could indicate a potential threat to space operations. Space-based Intelligence Community assets contribute intelligence, surveillance, and reconnaissance sensor data to support identification and characterization of threats, and to support Guardians in protecting and defending US, allied, and partner space capabilities.

Optimal Employment of Sensors and Data Sources

Historically, SSN planning and employment used a tasked-track methodology to optimize sensor track capability for maintaining custody of known objects in orbit. This means sensors were specifically tasked to track known objects to maintain custody of cataloged items and conduct space traffic management. Occasional search campaigns would be conducted to add new objects to the catalog but otherwise discovery of new objects typically occurred during tracking of a known object. The Space Force is moving to a search-based, task-as-required methodology that establishes a base of regular periodic searches to prevent, or at least minimize, surprise by hostile actors. The regular searches are supplemented by taskings to track specific known spacecraft or objects necessary to support missions requiring more accurate element sets or more persistent monitoring of non-cooperative spacecraft (i.e., spacecraft whose owner or operator does not provide notice of upcoming launches or maneuvers).

Sensors are assigned to search or tasked to perform track roles based on inherent system capabilities and current mission needs. Robust custody of priority space objects is maximized by using a variety of sensors. While some objects can only be observed by a single sensor, ideal custody includes sensor type and geographic diversity. For example, a radar can provide high quality range and rate data while optical sensors can provide photometric and angular accuracy.

a. **Missed Collection Opportunities**. For SSA purposes, it is important to understand situations that may result in a sensor failing to successfully complete a scheduled collection on a resident space object. Several factors can result in the failure to complete a successful collection and determine appropriate follow-on actions concerning the collection target. These factors include:

1. The sensor did not schedule the collection due to a higher-priority tasking.

2. The sensor did not schedule the collection due to weather or equipment issues.

3. The object maneuvered prior to the scheduled collection.

4. The sensor used an incorrect or outdated element set.

5. The scheduled collection was a low-probability pass (e.g., near the horizon) or the signal from the object was near or below the level of signals from other sources (i.e., noise floor) and therefore, was not "visible" to the sensor. An example of a low-probability pass is one where the target object is near the horizon resulting in a radar signal subject to atmospheric absorption losses.

Missed passes on suspicious space objects should prompt immediate action to search for them (missed pass report / no-show report to the C2 agency, and immediate voice tasking of the next several sensors in view) and possibly issue warnings if it was a high probability pass. These types of actions could result in wasted time and resources if it was a low probability pass.

b. **Capabilities of Link and Terrestrial Segments for SDA.** In addition to awareness of spacecraft and other resident space objects, the SDA mission also requires awareness of the terrestrial and link segments of space systems for mission accomplishment. Understanding the electromagnetic spectrum operations environment enables good stewardship of the space domain and the link segment by conducting frequency conjunction analysis to avoid satellite telemetry, tracking, and commanding issues. The following considerations capture what Guardians should understand in the context of SDA of these segments.

Comprehensive link segment awareness of space systems—friendly, neutral, and adversary—should include, but is not limited to, information pertaining to communication links and frequencies; broadcast signals; and uplink, downlink, and crosslink capabilities. Guardians can gain SDA of the link segment in multiple ways. For example, Guardians can gain awareness through lost-link reports from operators, indicating electromagnetic interference in the link segment. Electromagnetic warfare assets can monitor the electromagnetic spectrum to detect and characterize electromagnetic interference. Operators and cyberspace capabilities can detect, observe, and report indications of electromagnetic interference or issues with link integrity through network monitoring (e.g., data packet loss). The Intelligence Community and mission partners also have capabilities to monitor the link segment.

Complete terrestrial segment awareness of space systems should consist of, but is not limited to, information pertaining to launch sites and capabilities; telemetry, tracking, and commanding stations; personnel; communication nodes; space surveillance systems and capabilities; space dependencies; and status of systems. Guardians can gain SDA of the terrestrial segment through health and status reports of the ground architecture from operators. Operators and cyberspace capabilities can provide awareness of the health of networks connecting terrestrial nodes. Units providing security of the terrestrial segment (e.g., Air Force Security Forces, joint force security units) provide awareness of the terrestrial segment through force protection reporting. The Intelligence Community and mission partners also have capabilities to monitor the terrestrial segment and assess adversary actions including launch preparations, capability development and testing, identification of weapon systems with the ability to threaten space operations, and collaboration with other state and non-state actors. Guardians should exploit those capabilities to gain awareness of enemy and adversary actions and capabilities in the terrestrial segment.

SDA Mission Data

SDA presents a significant data challenge because of the large scope and availability of data required. Space Force SDA data systems should enable information mobility across multiple security classification levels.[7] SDA data considerations include:

a. **Volume.** Collecting, synthesizing, fusing, and understanding large volumes of data from various sources ensures maximum domain awareness. While fusing data collected by all current SSN sensors is a challenge for the Service, future sensors should be designed to enable subsequent data authentication and fusion.

b. **Variety.** Effective SDA requires diverse types and sources of data, including DoD sensors as well as academic, commercial, Intelligence Community, allied, and partner sources. For example, SSA relies on the fusion of space object tracking data of varying quality and sources to achieve the best awareness of the space terrain. Beyond SSA data, the full scope of SDA data sources and types, and the processing of that information, is a key challenge in the SDA mission area. The Space Force may not control the format or structure of all inputs due to requirements for data from across the Intelligence Community and other partner organizations. Guardians should be capable of fusing data and information from all available and applicable sources to provide timely and actionable SDA.

c. **Timeliness.** SDA signals and data pass through all mediums. The various mediums affect speed of the data transmission. Understanding this effect and optimizing the rate of data transfer could be critical to meeting the needs of a particular mission.

d. **Veracity.** Comparing data from multiple sources improves the accuracy of the resulting product. However, Guardians contend with data of varying levels of validation, fidelity, and quality. They need to account for data quality and threat of military deception to gain understanding with higher fidelity.

e. **Priority.** Prioritizing the criticality of any SDA information is key to current SDA operations. SDA data is used for a variety of operational functions, and the criticality of different types of information will be dependent on the given function or context. Guardians need to understand the mission needs so they can prioritize, integrate, and exploit available data in a meaningful, interpretable, and timely manner.

SDA operations depend on an accurate catalog of objects to identify new items in orbit or changes to items already tracked to identify potential hazards or threats to operational spacecraft. This list provides examples of key SDA activities that provide mission assurance:

a. **Environmental Monitoring.** Environmental monitoring includes collection in the natural environment, to include solar activity, geomagnetic activity, naturally occurring space objects, terrestrial weather effects, and naturally occurring electromagnetic interference, which pose hazards to US, allied, and partner space capabilities. Awareness of these hazards allows for mitigation actions or informs assessments about the possible source of issues with an asset.

b. **Conjunction Assessment.** Conjunction assessment is the process for determining the point and time of closest approach of two tracked orbiting objects. This includes matching active spacecraft against the spacecraft catalog several times per day to identify and predict close approaches. This information is essential for protection of high-value assets or identification of evolving threats. If a close approach meets emergency reportable criteria, the spacecraft's owner/operator should be notified so they can determine the appropriate action to avoid a collision.

   1. **Launch Conjunction Assessment.** Launch conjunction assessment identifies potential conjunctions that may result in a collision between objects being launched and objects already on orbit. The launch conjunction assessment screening results identify periods during the launch window which may put the booster or payload at increased risk for collision.

   2. **Collision Avoidance Maneuvers.** Spacecraft owner or operators make collision avoidance maneuver decisions in response to conjunction assessments. It is the responsibility of the spacecraft owner/operator to respond, but United States Space Command (USSPACECOM) supports collision avoidance efforts by screening predictive ephemeris data. The Space Force organizes, trains, and equips the forces and capabilities that enable USSPACECOM to screen collision avoidance data effectively.

c. **Deorbit and Reentry Support.** Deorbit and reentry support includes reentry assessment, reentry confirmation, and assistance in deorbit operations. Reentry assessments are predictions of the time and location where an uncontrolled object will reenter the atmosphere, not where the object will impact the ground. Deorbit and reentry support contributes to termination of missions that may be associated with the object deorbiting/reentering, accurate tracking and cataloging of the object, as well as assessing and predicting hazards associated with deorbiting and reentering. Deorbit and reentry support is critical to provide warning to terrestrial forces of impact or debris that may survive reentry. It is also part of shared warning to ensure a reentry is not perceived as a missile or orbital bombardment system attack.

Chapter 4: Organizations

Organizations across the Space Force and the broader space community collaborate to integrate intelligence, surveillance, and reconnaissance information; SSA data; and non-traditional data to develop reliable SDA. These different forms of information analyzed together to broaden the scope of knowledge from being reactive and maintaining a catalog of objects on orbit, to predictive SDA that considers missions, intentions, and system capabilities and equips decision makers with the near real-time information necessary to act within an adversary's decision cycle.

Space Operations Command (SpOC)

SpOC is the Space Force's field command responsible for generating, presenting, and sustaining combat-ready intelligence, cyber, space, and combat support forces.

Guardians assigned to the deltas under SpOC operate a wide range of terrestrial and on-orbit assets to collect space observation data. These dedicated and collateral sensors provide electro-optical and radar observations, space object identification data across the electromagnetic spectrum, nuclear detonation detection data, infrared launch detection data, and space-based imagery. Collateral sensors operated by the Space Force and operationally controlled by USSPACECOM provide space surveillance support as a secondary mission. Data from these sensors supports SDA development and generates warning intelligence and other products in support of external users and internal Space Force operations. SpOC prepares and presents assigned and attached forces to execute SDA operations as part of the joint force to deter aggression and, if necessary, protect and defend the US and our allies from attack in, from, and to space. SpOC space deltas are either directly responsible for, contribute to, or rely on SDA as presented forces in execution of the following mission areas:

a. Monitoring activity in, from, and to space.

b. Tasking, tracking, identifying, characterizing, and responding to events by employing the SSN and supporting space deltas.

c. Employing space electromagnetic warfare capabilities to protect and defend US and allied global operations.

d. Defending US space capabilities by protecting critical satellite communications links with a near-global capability to detect, characterize, geolocate, and report sources of electromagnetic interference on US military and commercial spacecraft.

e. Providing immediate warning of harmful or hostile activity in, from, or to the space domain.

f. Tracking and characterizing objects in Earth orbit and allied, commercial, or non-cooperative launches.

g. Maintaining global awareness of operational environments and the status of space forces, enable data-driven decisions by the operational commander, and task the operational commander's forces.

    h. Providing immediate threat warning assessment of cyberspace activity in, from, or to the space domain.

    i. Providing critical, time-sensitive, and actionable intelligence for space domain operations to allow for the detection, characterization, and targeting of adversary space capabilities.

    j. Conducting space-based battlespace characterization operations.

    k. Commanding space-based SDA spacecraft to collect and disseminate decision-quality information across the spectrum of conflict.

    l. Providing foundational, scientific, and technical intelligence to inform senior policy makers, service and national acquisitions, and military operations.

    m. Fusing Intelligence Community sensor data into DoD networks and decision making.

United States Space Command (USSPACECOM)

USSPACECOM executes the operational C2 of assigned and attached SDA forces to achieve theater and global objectives. It continuously coordinates, plans, integrates, synchronizes, and executes global space operations, providing tailored space effects on demand to support combatant commanders and accomplishing national security objectives. USSPACECOM is also responsible for, in unified action with mission partners, deterring aggression, defending capabilities, and defeating adversaries throughout the competition continuum to maintain space superiority in the USSPACECOM area of responsibility.

Contributing Sensors

Numerous organizations outside the Space Force operate radars or other sensors that contribute space surveillance support to the SSN upon request. Examples of these organizations include Massachusetts Institute of Technology Lincoln Laboratory through the Lincoln Space Surveillance Complex; research and development optical sensors at the Maui Space Surveillance Site on Mount Haleakala, Hawaii; radars operated by Army Space and Missile Defense Command at the Reagan Test Site on Kwajalein Atoll; and the Sapphire Optical Satellite operated by the Canadian Department of National Defence.

Data Sharing (Cooperation)

Data sharing is key to space security because it improves SDA for all parties. Effective sharing requires development of key partnerships among the DoD, the Intelligence Community, intragovernmental agencies, international bodies, civil agencies, allies, partners, academic, and commercial entities. This can even include competitors and adversaries in the appropriate circumstances where potential areas of shared interest exist such as space debris mitigation or collision avoidance. Additionally, data sharing agreements enable the US to operate from geographic locations necessary for achieving robust SDA. Effective data sharing is dependent upon timely and secure communication channels and encryption protocols to protect sensitive information while enabling rapid decision making and response to potential threats. In addition to data sharing, collaboration on analytical processes and methods helps promote greater interoperability with partners and allies, many of whom have already integrated with US combat capability in other domains. This requires clear data sharing agreements that protect intellectual property rights and sensitive information and ensure compliance with applicable laws and policy requirements.

a. **Coalition Allies and Partners.** Strengthening relationships with allies and partners, by sharing intelligence data and analytical processes and methods, is critical to promoting space security. The Space Force continually reevaluates data sharing and collaboration agreements, and security paradigms that often limit interoperability with allies and partners in space.

   The Space Force's Unified Data Library, a cloud-based data environment, is a key component of the Space Force's digital architecture. The Space Force is working to increase information sharing with allies by developing an allied exchange environment, which will connect other countries to the Unified Data Library.[7]

b. **Academic, Civil, and Commercial Organizations.** The Space Force integrates data, analysis capabilities, and support from other Services, academia, civil, and commercial space entities. The United States Department of Commerce is the civil agency interface for space traffic management, collision avoidance and basic SSA. Remote sensing, surveillance and reconnaissance, geolocation, targeting, radio frequency emissions, and imagery from academic, civil, and commercial space entities enhance DoD and the Intelligence Community's space-based intelligence data. As part of these relationships the Space Force continually reevaluates data sharing and collaboration agreements, and security paradigms that often limit sharing.

c. **Interoperable Data Formats and Common Messages.** International partnerships facilitate SDA in support of allied/combined operations and within the constraints of US national security requirements. When possible, data sources (i.e., sensors) and processing centers should use common data formats to enable dissemination and processing. When not possible, the data source should communicate to a system that can translate the data to a common data format used by the data processing sensors. Similarly, C2 centers should use common messages that allied, partner, commercial, and academic sensors can use in addition to military sensors.

APPENDIX A: Acronyms, Abbreviations, and Initialisms

ASAT	antisatellite
C2	command and control
DoD	Department of Defense
GEO	geosynchronous Earth orbit
LEO	low Earth orbit
NASA	National Aeronautics and Space Administration
SDA	space domain awareness
SpOC	Space Operations Command
SSA	Space Situational Awareness
SSN	Space Surveillance Network
STARCOM	Space Training and Readiness Command
US	United States
USSF	United States Space Force
USSPACECOM	United States Space Command

APPENDIX B: Glossary

Cislunar regime — the combined Earth-Moon two body gravitational system. The cislunar regime is nested within the **solar regime**. (Space Capstone Publication, *Spacepower*)

Conjunction assessment — Analysis of observation data when on-orbit asset behavior and orbit characteristics indicate the potential of collision or a close approach of two on-orbit objects.

Debris – For space, refers to any spacecraft or artificial satellite (e.g., a rocket body) in orbit that no longer serves a useful purpose. (Space Capstone Publication, *Spacepower*)

Disposal orbit – An orbit where satellites are placed near the end of their operational life to reduce the probability of colliding with operational spacecraft and generating space debris.

Drag – Within terrestrial domains, atmospheric density and pressure that resist all forms of motion by generating viscous friction. (Space Capstone Publication, *Spacepower*)

Electromagnetic interference — Any electromagnetic disturbance, induced intentionally or unintentionally, that interrupts, obstructs, or otherwise degrades or limits the effective performance of electromagnetic spectrum-dependent systems and electrical equipment. (Joint Publication 3-85)

Electromagnetic jamming — The deliberate radiation, reradiation, or reflection of electromagnetic energy for the purpose of preventing or reducing an enemy's effective use of the electromagnetic spectrum, with the intent of degrading or neutralizing the enemy's combat capability. (JP 3-85)

Electromagnetic warfare — Military action involving the use of electromagnetic and directed energy to control the electromagnetic spectrum or to attack the enemy. Also called **EW.** (JP 3-85)

Geomagnetic storm — A major disturbance of Earth's magnetosphere that occurs when there is a very efficient exchange of energy from the solar wind into the space environment surrounding Earth.[4] (NOAA)

Geosynchronous Earth orbit — An orbit synchronized to the Earth's rotation, orbiting at the same rate at which the Earth rotates upon its axis. Satellites in this orbit have an altitude of approximately 23,000 miles above the Earth's surface and create a figure eight ground trace over the ground. Also called **GEO.**

Information mobility – Provides timely, rapid, and reliable collection and transportation of data across the range of military operations in support of tactical, operational, and strategic decision making. (Space Capstone Publication, *Spacepower*)

Intelligence Community — All departments or agencies of a government concerned with intelligence activity, in either an oversight, managerial, support, or participatory role. Also called **IC.** (JP 2-0)

Intelligence, surveillance, and reconnaissance — 1. An integrated operations and intelligence activity that synchronizes and integrates the planning and operation of sensors; assets; and processing, exploitation, and dissemination systems in direct support of current and future operations. 2. The organizations or assets conducting such activities. Also called **ISR**. (JP 2-0)

Low Earth orbit — Orbits that are at a height of approximately 1,000 miles or less above the surface of the Earth and average time to orbit the Earth of approximately 90-100 minutes. Also called **LEO**.

Mission assurance — A process to protect or ensure the continued function and resilience of capabilities and assets, including personnel, equipment, facilities, networks, information and information systems, infrastructure, and supply chains, critical to the execution of DoD mission-essential functions. (JP 3-26)

Node – For space operations, an element of the space architecture capable of creating, processing, receiving, or transmitting data. (Space Capstone Publication, *Spacepower*)

Operations in the information environment — Military actions involving the integrated employment of multiple information forces to affect drivers of behavior. Also called **OIE.** (JP 3-04)

Orbital regime – a region in space associated with a dominant gravitational system capable of capturing the orbit of other objects. See also **geocentric regime**, **cislunar regime**, and **solar regime**. (Space Capstone Publication, *Spacepower*)

Space access and sustainment – 1. Knowledge of processes, support, and logistics required to maintain and prolong operations in the space domain. 2. Ability to resource, apply, and leverage spacepower in, from, and to the space domain. (Space Capstone Publication, *Spacepower*)

Space battle management – 1. Knowledge of how to orient to the space domain and skill in making decisions to preserve mission, deny adversary access, and ultimately ensure mission accomplishment. 2. Ability to identify hostile actions and entities, conduct combat identification, target, and direct action in response to an evolving threat environment. (Space Capstone Publication, *Spacepower*)

Space domain — The area above the altitude where atmospheric effects on airborne objects become negligible. (JP 3-14)

Space domain awareness — The timely, relevant, and actionable understanding of the operational environment that allows military forces to plan, integrate, execute, and assess space operations. Also called **SDA**. (JP 3-14)

Space environment — The environment corresponding to the space domain, where electromagnetic radiation, charged particles, and electric and magnetic fields are the dominant physical influences, and that encompasses the Earth's ionosphere and magnetosphere, interplanetary space, and the solar atmosphere. (JP 3-59)

Space operations — The employment of space forces in, to, or from space to achieve objectives. (JP 3-14)

Space situational awareness — The requisite foundational, current, and predictive knowledge and characterization of space orbital objects and the space domain. Also called **SSA**. (JP 3-14)

Space superiority — The degree of control in the space domain of one force over another that permits freedom of access and action without prohibitive interference. (JP 3-14)

Space Surveillance Network — A combination of optical and radar sensors used to detect, track, identify, and catalog objects orbiting the Earth. Also called **SSN**.

spacecraft – An object which has been engineered to be controlled and deliberately employed to perform a useful purpose while traveling in, from, and to the space domain. (Space Capstone Publication, *Spacepower*)

Threat warning and assessment — The evaluation of potential or actual attacks in all domains, space weather effects, and space system anomalies and the subsequent notification to decision makers.

Uncorrelated track — An object observed in space where the data indicates an object not correlated to previously known objects in the space surveillance catalog.

APPENDIX C: References

1. Albon, C. (2022c, February 13). *Air Force Research Lab building momentum on cislunar projects*. C4ISRNet. Retrieved January 31, 2023, from https://www.c4isrnet.com

2. ARES | Orbital Debris Program Office | Debris Modeling. (n.d.). NASA. Retrieved July 12, 2023, from https://www.orbitaldebris.jsc.nasa.gov/quarterly-news

3. Defense Intelligence Agency. (2022, March). *Challenges to security in space: space reliance in an era of competition and expansion*. Retrieved January 31, 2023, from https://www.dia.mil

4. Fang, T.-W., Kubaryk, A., Goldstein, D., Li, Z., Fuller-Rowell, T., Millward, G., et al. (2022, November 2). *Space weather environment during the SpaceX Starlink satellite loss in February 2022*. Retrieved January 31, 2023, from https://agupubs.onlinelibrary.wiley.com

5. Bingen, K., Johnson, K., Young, M., Raymond, J. (2023, April 14). *Space Threat Assessment 2023*. Center for Strategic & International Studies. Retrieved July 13, 2023, from https://www.csis.org

6. Malik, T. (2022, February 9). *SpaceX says a geomagnetic storm just doomed 40 Starlink internet satellites.* Space.Com. Retrieved January 31, 2023, from https://www.space.com

7. U.S. Space Force. (2023, May). *U.S. Space Force Vision for Space Domain Awareness.*

APPENDIX D: Strategic Guidance, Policy, and Doctrine

1. **National Security Strategy, October 2022** – Outlines how the United States will advance our vital interests and pursue a free, open, prosperous, and secure world. We will leverage all elements of our national power to outcompete our strategic competitors; tackle shared challenges; and shape the rules of the road. The Strategy is rooted in our national interests: to protect the security of the American people, to expand economic opportunity, and to realize and defend the democratic values at the heart of the American way of life.

2. **2022 National Defense Strategy of the United States of America, 27 October 2022** – Details the Department's path forward into a decisive decade—from helping to protect the American people, to promoting global security, to seizing new strategic opportunities, and to realizing and defending our democratic values. The National Defense Strategy directs the Department to act urgently to sustain and strengthen US deterrence, with the People's Republic of China as the pacing challenge for the Department. The National Defense Strategy further explains how we will collaborate with our North Atlantic Treaty Organization allies and partners to reinforce robust deterrence in the face of Russian aggression while mitigating and protecting against threats from North Korea, Iran, violent extremist organizations, and transboundary challenges such as climate change.

3. **Defense Space Strategy Summary, June 2020** – Identifies how DoD will advance spacepower to enable the Department to compete, deter, and win in a complex security environment characterized by great power competition.

4. **United States Space Priorities Framework, December 2021** – Focuses on advancing and synchronizing our civil, commercial, and national security space activities, and adds emphasis in support of the (Biden) administration's agenda, including promoting peaceful exploration of space and reducing the risk of miscalculation or conflict in space; addressing the climate crisis; and enhancing science, technology, engineering, and math education.

5. **National Space Policy of the United States of America, 9 December 2020** – Sets out the nation's commitment to leading in the responsible and constructive use of space, promoting a robust commercial space industry, returning Americans to the Moon, and preparing for Mars, leading in exploration, and defending United States and allied interests in space.

6. **DoD Directive 3100.10, *Space Policy*, 30 August 2022** – Establishes policy and assigns responsibilities for DoD space-related activities in accordance with the National Space Policy, the U.S. Space Priorities Framework, the National Defense Strategy, the Defense Space Strategy, and U.S. Law, including Titles 10, 50, and 51, United States Code.

7. **Space Capstone Publication, 10 August 2020** – The capstone doctrine for the United States Space Force and represents the Service's first articulation of an independent theory of spacepower. This publication answers why spacepower is vital for our Nation, how military spacepower is employed, who military space forces are, and what military space forces value.

8. **Chief of Space Operations' Planning Guidance, 2020** – Provides foundational direction for the Space Force to advance National and DoD strategic objectives. This authoritative

Service-level planning guidance supersedes previous guidance and provides the context and outline for our new Service design.

9. **Joint Publication 3-14, *Joint Space Operations*** – This publication provides fundamental principles and guidance to plan, execute, and assess joint space operations. It sets forth joint doctrine to govern the activities and performance of the Armed Forces of the United States in joint operations, and it provides considerations for military interaction with governmental and nongovernmental agencies, multinational forces, and other interorganizational partners. It provides military guidance for the exercise of authority by combatant commanders and other joint force commanders and prescribes joint doctrine for operations and training.

10. **Space Doctrine Publication 2-0, *Intelligence*** – Presents Space Force intelligence operations to support the freedom to operate in, from, and to space. It introduces intelligence as part of military spacepower; discusses the role of intelligence and its integration with other spacepower disciplines; presents the application of joint intelligence disciplines to space intelligence; describes the intelligence process, intelligence collection, and collection authorities; and presents Space Force organizations supporting intelligence.

11. **Space Doctrine Publication 3-0, *Operations*** – Presents Space Force delivery of spacepower as an independent option for national and joint leadership, and as a part of unified action under a joint force commander. Provides an overview of military space operations and their contributions to joint all-domain operations, discusses the space operational environment, details the operational concept of spacepower, and discusses the Space Force structure and presentation of space forces to the joint force.

12. **Unified Command Plan** – An executive branch document prepared by the chairman of the Joint Chiefs of Staff that assigns missions; planning, training, and operational responsibilities; and geographic areas of responsibility to combatant commands.

13. **Department of Defense Electromagnetic Spectrum Superiority Strategy, October 2020** – Addresses how DoD will: develop superior electromagnetic spectrum capabilities; evolve to an agile, fully integrated electromagnetic spectrum infrastructure; pursue total force electromagnetic spectrum readiness; secure enduring partnerships for electromagnetic spectrum advantage; and establish effective electromagnetic spectrum governance to support strategic and operational objectives. Investment in these areas will speed decision-quality information to the warfighter, establish effective electromagnetic battle management, enable electromagnetic spectrum sharing with commercial partners, advance electromagnetic spectrum warfighting capabilities, and ensure our forces maintain electromagnetic spectrum superiority.

14. **National Cislunar Science and Technology Strategy** – Guides the actions of the DoD and US government at large in advancing science and technology goals in Cislunar space. This strategy identifies key objectives to support research and development to enable long-term growth in cislunar space; expand international science and technology cooperation in cislunar space; extend US SSA capabilities into cislunar space; and implement cislunar communications and positioning, navigation, and timing capabilities with scalable and interoperable approaches.

15. [National Preparedness Strategy & Action Plan for Near-Earth Object Hazards and Planetary Defense](#) – Guides DoD and other US government agencies and departments in developing capabilities and technologies necessary to enhance hazardous object detection and mitigation.

APPENDIX E: Expanding SDA Requirements

As the likelihood of threats emanating from the cislunar regime beyond GEO transitions from part of a far-off future to an issue we expect to contend with in the foreseeable future, the urgency to obtain and maintain SDA in that regime increases.

Figure 6. Air Force Research Lab depiction of Oracle - Prime[1]

Most current space operations take place in the geocentric orbital regime. In the geocentric regime, Earth's gravity dominates, and objects follow orbital trajectories relative to Earth. Increased operations beyond the GEO belt dictate expanding SDA requirements. Current sensor capabilities will find that the vastness of space between the Earth and the Moon, and around the Moon, creates challenging conditions for search, custody, and collection operations in support of joint forces. New sensors of all types, particularly space-based, will be required to develop and maintain SDA beyond GEO.

Space Force support of NASA's Artemis 1 mission in late 2022 provided an opportunity to test tracking capabilities in cislunar space beyond GEO. Space Force units collaborated to maintain custody of objects beyond GEO, where space object tracking becomes much more complicated due to the increased gravitational effects of the Sun and Moon. The Space Force also functioned as a liaison by sharing information, data points, and lessons learned among the DoD, civil, commercial, academic, and other government partners. The requirement to further develop the capabilities necessary to develop and maintain SDA of cislunar space will undoubtedly continue to expand as friendly, neutral, and adversarial actors increase their focus on conducting operations beyond GEO.

APPENDIX F: Cornerstone Responsibilities, Core Competencies and Spacepower Disciplines

Cornerstone responsibilities. Military space forces conduct prompt and sustained space operations, accomplishing three cornerstone responsibilities. Taken together, these cornerstone responsibilities define the vital contributions of military spacepower. (Space Capstone Publication, *Spacepower*)

Preserve freedom of action. Unfettered access to and freedom to operate in space is a vital national interest; it is the ability to accomplish all four components of national power – diplomatic, information, military, and economic – of a nation's implicit or explicit space strategy. Military space forces fundamentally exist to protect, defend, and preserve this freedom of action. (Space Capstone Publication, *Spacepower*)

Enable joint lethality and effectiveness. Space capabilities strengthen operations in the other domains of warfare and reinforce every joint function – the United States does not project or employ power without space. At the same time, military space forces must rely on military operations in the other domains to protect and defend space freedom of action. Military space forces operate as part of the closely integrated joint force across the entire conflict continuum in support of the full range of military operations. (Space Capstone Publication, *Spacepower*)

Provide independent options. Providing the ability to achieve strategic effects independently is a central tenet of military spacepower. In this capacity, military spacepower is more than an adjunct to landpower, seapower, airpower, and cyberpower. Across the conflict continuum, military spacepower provides national leadership with independent military options that advance the Nation's prosperity and security. Military space forces achieve national objectives by projecting power in, from, to space. (Space Capstone Publication, *Spacepower*)

Core competencies. The United States Space Force executes five core competencies. These core competencies represent the broad portfolio of capabilities military space forces need to provide successfully or efficiently to the Nation. (Space Capstone Publication, *Spacepower*)

Space security. Space security establishes and promotes stable conditions for the safe and secure access to space activities for civil, commercial, Intelligence Community, and multinational partners. (Space Capstone Publication, *Spacepower*)

Combat power projection. Combat power projection integrates defensive and offensive operations to maintain a desired level of freedom of action relative to an adversary. Combat Power Projection in concert with other competencies enhances freedom of action by deterring aggression or compelling an adversary to change behavior. (Space Capstone Publication, *Spacepower)*

Space mobility and logistics. Space mobility and logistics enables movement and support of military equipment and personnel in the space domain, from the space domain back to Earth, and to the space domain. (Space Capstone Publication, *Spacepower*)

Information mobility. Information mobility provides timely, rapid, and reliable collection and transportation of data across the range of military operations in support of tactical, operational, and strategic decision making. (Space Capstone Publication, *Spacepower*)

Space domain awareness. The timely, relevant, and actionable understanding of the operational environment that allows military forces to plan, integrate, execute, and assess space operations. (Joint Publication 3-14, *Joint Space Operations*)

Spacepower disciplines. The seven spacepower disciplines are necessary components of military spacepower theory. These disciplines are the skills the United States Space Force needs when developing its personnel to become the masters of space warfare. (Space Capstone Publication, *Spacepower*)

Orbital warfare. Knowledge of orbital maneuver as well as offensive and defensive fires to preserve freedom of access to the domain. Skill to ensure United States and coalition space forces can continue to provide capability to the joint force while denying that same advantage to the adversary. (Space Capstone Publication, *Spacepower*)

Space electromagnetic warfare. Knowledge of spectrum awareness, maneuver within the spectrum, and non-kinetic fires within the spectrum to deny adversary use of vital links. Skill to manipulate physical access to communication pathways and awareness of how those pathways contribute to adversary advantage. (Space Capstone Publication, *Spacepower*)

Space battle management. Knowledge of how to orient to the space domain and skill in making decisions to preserve mission, deny adversary access, and ultimately ensure mission accomplishment. Ability to identify hostile actions and entities, conduct combat identification, target, and direct action in response to an evolving threat environment. (Space Capstone Publication, *Spacepower*)

Space access and sustainment. Knowledge of processes, support, and logistics required to maintain and prolong operations in the space domain. Ability to resource, apply, and leverage spacepower in, from, and to the space domain. (Space Capstone Publication, *Spacepower*)

Military intelligence. Knowledge to conduct intelligence-led, threat-focused operations based on the insights. Ability to leverage the broader Intelligence Community to ensure military spacepower has the ISR capabilities needed to defend the space domain. (Space Capstone Publication, *Spacepower*)

Engineering and acquisition. Knowledge that ensures military spacepower has the best capabilities in the world to defend the space domain. Ability to form science, technology, and acquisition partnerships with other national security space organizations, commercial entities, allies, and academia to ensure the warfighters are properly equipped. (Space Capstone Publication, *Spacepower*)

Cyber operations. Knowledge to defend the global networks upon which military spacepower is vitally dependent. Ability to employ cyber security and cyber defense of critical space networks and systems. Skill to employ future offensive capabilities. (Space Capstone Publication, *Spacepower*)